The Analysis of
Tidal Stream Power

The Analysis of
Tidal Stream Power

JACK HARDISTY
The University of Hull, Kingston-upon-Hull, UK

WILEY-BLACKWELL

A John Wiley & Sons, Ltd., Publication

This edition first published 2009
© 2009 by John Wiley & Sons Ltd.

Wiley-Blackwell is an imprint of John Wiley & Sons, formed by the merger of
Wiley's global Scientific, Technical and Medical business with Blackwell Publishing.

Registered office
John Wiley & Sons Ltd, The Atrium, Southern Gate, Chichester, West Sussex,
PO19 8SQ, UK

Other Editorial Offices:
9600 Garsington Road, Oxford, OX4 2DQ, UK
111 River Street, Hoboken, NJ 07030-5774, USA

For details of our global editorial offices, for customer services and for information about
how to apply for permission to reuse the copyright material in this book please see our
website at www.wiley.com/wiley-blackwell

Library of Congress Cataloging-in-Publication Data

Hardisty, J. (Jack), 1955–
  The analysis of tidal stream power / Jack Hardisty
    p. cm.
  Includes bibliographical references and index.
  ISBN 978-0-470-72451-4
  1. Tidal power-plants. 2. Tidal power. 3. Tide-waters. I. Title
  TK1081.H342 2009
  621.31′2134—dc22

                                              2008044506

A catalogue record for this book is available from the British Library.

ISBN 978-0-470-72451-4 (H/B)

Set in 10/12pt Sabon by Integra Software Services Pvt. Ltd, Pondicherry, India
Printed in Great Britain by CPI Antony Rowe, Chippenham, Wiltshire.

First [Impression]   2009

Contents

Part II   Practice

Preface

I have been engaged in marine environmental research throughout my career, but came latterly to the subject of Renewable Energy. My first contact involved an analysis of the potential for tidal power around the United Kingdom. This appeared straightforward, and 18 sites were identified (most of which are detailed now in Chapter 7) from British Admiralty tidal diamonds, and their hydraulic powers were calculated. The draft report, however, missed the Pentland Firth because there are no tidal diamonds in this very high-current regime. Therefore, when the *DTI Tidal Resource Atlas* was published a few weeks later, it was apparent that the methodology was correct (basic physics) but the results were not. There was much more to this renewable energy business than initially met the eye, and I became addicted. A whole new world of fascinating research problems opened up as we worked for some of the major players such as Lunar Energy, ITPower, Pulse Generation, and Neptune. I strove to maintain academic rigour and peer review in a fast-moving field with a harshly commercial environment. The result is this book, which attempts to set down, for the first time, the fundamental physics behind tidal stream power alongside a global analysis of its distribution and potential.

I have been very fortunate to work with some of the best British practitioners. Thanks are due to Simon Meade at Lunar Energy, Jamie O'Nians and Huw Traylor from IT Power, Pete Stratford (then) from BMT Renewables, Marc Paish from Pulse Generation, Glenn Aitken, Andrew Laver, and Nigel Petrie at Neptune Renewable Energy, and Nathalie Stephenson from Atkins Global. I have also engaged with many industrial and business people including Graham Bilaney, formerly at Dunstons, Stuart Reasbeck at IMT Marine, Ian Mitchell at Ormston, MMS Shiprepairers, David Brown Gearboxes, and the electrical engineers at Sprint and Brook Compton. Much has been learnt from these specialists.

Thanks are also due to many academic colleagues, and to the students who quickly and willingly took up undergraduate and graduate dissertations in Renewable Energy. There is a growing group in Hull University who have taken Research Masters and Doctoral programmes on some of the problems detailed in these pages, including Emma Toulson and the MRes students Tom Smith, Chris Smith, and Paul Jensen. It is an old aphorism, but no less valid that: teaching remains the best way of learning. The colleagues with whom I have discussed much, and among whom we have developed our University's Renewable Energy centre of excellence, include Stuart McLelland, Brendan Murphy, David Calvert, and Professors Lynne Frostick, Tom Coulthard, and Mike Elliott. John Garner drew many of the diagrams herein.

In addition, and for completely different reasons, much of this book was written on the Haemodialysis Unit at Hull Royal Infirmary, and my thanks are due to my consultant, David Eadington, and to the ward staff, particularly Sue Smith and Rita Soames. Writing was initially interrupted and later enhanced by very useful stays in the excellent Renal Transplant Unit at St James' Hospital; thanks are due to the staff there and in, particular, to Mr Ahmed and the team on Ward 59.

Finally, there is my indulgent family; I gratefully acknowledge the help and support of Paul, Tor, Lexie, Lizzie, Annette, and, in particular, my son James for always being there. Last, but by no means least, this book is for Sarah.

Jack Hardisty
East Yorkshire
July 2008

Website contents

www.wileyeurope.com/college/hardisty

Origin of online images

This Table is available in digital format on the book's website
www.wileyeurope.com/college/hardisty

1.1   http://en.wikipedia.org/wiki/Aristotle
1.2   http://www-groups.dcs.st-and.ac.uk/~history/PictDisplay/Ptolmey.html
1.3   http://www-groups.dcs.st-and.ac.uk/~history/PictDisplay/Copernicus.html
1.4   http://www-groups.dcs.st-and.ac.uk/~history/PictDisplay/Brahe.html
1.5   http://zuserver2.star.ucl.ac.uk/~apod/apod/ap960831.html
1.7   http://zuserver2.star.ucl.ac.uk/~apod/apod/ap960830.html
1.8   http://math.usask.ca/conicsdemo/DEMO/extensions/images/descartes.jpeg
1.9   http://www.crystalinks.com/newton.html
1.13 http://content.scholastic.com/browse/article.jsp?id=5173
1.15 http://www.marxists.org/glossary/people/l/a.htm
1.16 http://snews.bnl.gov/popsci/spectroscope.html
1.17 http://www.ingenious.org.uk/See/Naturalworld/Oceanography/?pageNo=3&
      s=S1&viewby=images&
1.18 http://www.gavinrymill.com/flaybrick/doodson/
1.19 http://www.gavinrymill.com/flaybrick/doodson/
1.21 http://www.nature.com/nature/journal/v410/n6832/images/4101030aa.0.jpg
1.22 http://www.uh.edu/engines/epi2194.htm
1.23 http://www.lycee-fourneyron.fr/etablissement/benoit.html
1.24 http://www.todayinsci.com/9/9_18.htm
1.25 http://www.invent.org/hall_of_fame/293.html
1.26 http://www.kekenergia.hu/viz/kaplanfoto.jpg
2.1   http://tidesandcurrents.noaa.gov/index.shtml
2.3   http://en.wikipedia.org/wiki/Image:Oreynolds.jpg
2.5   http://www.todayinsci.com/11/11_28.htm
2.7   http://en.wikipedia.org/wiki/Image:Danielbernoulli.jpg
2.8   http://en.wikipedia.org/wiki/Theodore_von_K%C3%A1rm%C3%A1n
2.11 http://www-groups.dcs.st-and.ac.uk/~history/Mathematicians/Airy.html
3.3   http://en.wikipedia.org/wiki/Tide_mill
3.4   Left: http://www.lunarenergy.co.uk
      Right: http://www.cleancurrent.com
3.9   http://history.nasa.gov/SP-367/f38.htm
3.10 http://www.dopsys.com/loads.htm
3.11 http://www.search.com/reference/Airfoil
3.13 http://www.dti.gov.uk/files/file29969.pdf

Copyright acknowledgements

Symbols

a	acceleration, as in, for example, Newton's laws (m s^{-1}) (Section 1.4)
a_A	speed of constituent A in the Kelvin equation (Section 1.8) or in the harmonic current equation (Section 1.9)
a_K	length of the semi-minor axis in Kepler's first law (m) (Section 1.3)
A	cross-sectional area of the hydrofoil blade (m^2)
A_o	capture area of the device (m^2) (Section 3.4)
A_R	flow area at the rotor (m^2) (Section 3.5); duct cross-sectional area at the rotor or turbine (m^2)
b_K	length of the semi-major axis in Kepler's first law (m) (Section 1.3)
B	number of blades in the turbine or rotor
c	Airy wave phase celerity (m s^{-1}) (Section 2.9)
C_D	drag coefficient (Section 3.6)
C_{D100}	boundary layer drag coefficient (Section 2.7)
C_L	lift coefficient (Section 3.7)
C_P	power capacity factor of the device (Section 3.3)
d	distance used in Newton's analysis (m) (Section 1.4)
e	overall device efficiency (Section 3.3)
e_E	electrical efficiency (Section 3.3)
e_H	hydraulic efficiency of the ducting (Section 3.3)
e_R	rotor or turbine efficiency (Section 3.3)
f	factor reducing amplitude to year of prediction in the Kelvin equation (Section 1.8)
F	force, as in, for example, Newton's laws (N) (Section 1.4)
Fr	Froude number (Section 2.4)
Fz	Formzahl number (Section 5.5.2)
G	gravitational acceleration; universal gravitational constant, as in, for example, Newton's laws (kg) (Section 1.4)
h	height of tide in the Kelvin equation (m) (Section 1.8) or Airy wave (Section 2.9)
$h(q)$	height of the tide at q in Newton's analysis (m) (Section 1.4)
h_{max}	maximum height of the tide in Newton's analysis (m) (Section 1.4)
H	mean amplitude of any constituent A in the Kelvin equation (Section 1.8)
H_o	mean water level above datum in the Kelvin equation (Section 1.8)
I_T	turbulence intensity (Section 2.11)
k	wave number in the Airy expressions $= 2p/l$ (Section 2.8)

K	epoch of constituent A in the Kelvin equation (Section 1.8)
$K_E(t)$	kinetic energy per unit area in a tidal current (J m^{-2}) (Section 2.3)
K_z	coefficient of eddy viscosity in, for example, the Reynolds experiment (Section 2.4)
K_1	lunisolar diurnal constituent (Section 1.7)
K_2	lunisolar semi-diurnal constituent (Section 1.7)
L_2	smaller lunar elliptic semi-diurnal constituent (Section 1.7)
m	mass, as in, for example, Newton's laws (Section 1.4)
M	mass, as in, for example, Newton's laws (kg) (Section 1.4), or hydraulic power (Section 2.3)
M_E	mass of the Earth in, for example, Newton's analysis (kg) (Section 1.4)
M_f	lunar fortnightly constituent (Section 1.7)
M_m	lunar monthly constituent (Section 1.7)
M_M	mass of the moon in, for example, Newton's analysis (kg) (Section 1.4)
M_{U2}	principal lunar semi-diurnal current constituent (Section 5.5.3)
M_2	principal lunar semi-diurnal constituent (Section 1.7)
N	rate of revolution of the rotor (rev min^{-1}) (Section 3.6)
N_2	larger lunar elliptic semi-diurnal constituent (Section 1.7)
O_1	lunar diurnal constituent (Section 1.7)
p	pressure in, for example, the Bernoulli equation (N m^{-2}) (Section 2.6)
P	coefficient in Kepler's second law (Section 1.3)
$P_D(t)$	hydraulic power density (W m^{-2}) (Section 2.3)
P_E	electrical output power of the device (W) (Section 3.3)
P_H	hydraulic power across the capture area of the device (W) (Section 3.3)
P_I	installed power of the device (W) (Section 3.3)
P_R	hydraulic power at the rotor or turbine (W) (Section 3.3)
P_S	turbine shaft power (W) (Section 3.3)
P_1	solar diurnal constituent (Section 1.7)
Q_1	larger lunar elliptic diurnal constituent (Section1.7)
r	distance used in Newton's analysis (m) (Section 1.4)
R	separation of masses, as in, for example, Newton's laws (m) (Section 1.4)
R_F	radius of the point of action of the force on the blade (m)
R_R	radius of the rotor (m) (Section 3.6)
R_1	semi-major axis of planet 1 in Kepler's third law (m) (Section 1.3)
R_2	semi-major axis of planet 2 in Kepler's third law (m) (Section 1.3)
Re	Reynolds number (Section 2.4)
S	distance used in Newton's analysis (m) (Section 1.4)
S_{sa}	solar semi-annual constituent (Section 1.7)
S_{U2}	principal solar semi-diurnal constituent (Section 5.5.3)
S_2	principal solar semi-diurnal constituent (Section 1.7)
t	time, for example, in the Kelvin equation (s) (Section 1.8)
T	tidal period, for example, in the Airy wave (s) (Section 2.8)
T_S	period of rotation of the shaft (s) (Section 3.6)
T_{SR}	tip speed ratio (Section 3.6)
T_1	sidereal period of planet 1 in Kepler's third law (m) (Section 1.3)
T_2	sidereal period of planet 2 in Kepler's third law (m) (Section 1.3)

u	velocity in, for example, the Bernoulli equation ($\mathrm{m\ s^{-1}}$) (Section 2.6)
u_{100}	tidal-current speed at $100\,\mathrm{cm}$ above the bed ($\mathrm{m\ s^{-1}}$) (Section 2.7)
U	horizontal downstream current in turbulent decomposition ($\mathrm{m\ s^{-1}}$) (Section 2.11)
$U(t)$	speed of tidal current at any time t in the harmonic current equation (Section 1.9)
U_A	mean amplitude of any constituent A in the harmonic current equation (Section 1.9)
U_M	horizontal downstream mean current in turbulent decomposition ($\mathrm{m\ s^{-1}}$) (Section 2.11)
U_o	mean current speed (usually $U_o = 0$) in the harmonic current equation (Section 1.9)
U_R	flow velocity at the rotor ($\mathrm{m\ s^{-1}}$) (Section 3.5)
U'	horizontal downstream turbulent current component ($\mathrm{m\ s^{-1}}$) (Section 2.11)
U_*	friction velocity ($\mathrm{m\ s^{-1}}$) (Section 2.4)
U_0	tidal velocity at the entrance to the device ($\mathrm{m\ s^{-1}}$) (Section 3.5)
V	horizontal cross-stream current in turbulent decomposition ($\mathrm{m\ s^{-1}}$) (Section 2.11)
V_M	horizontal cross-stream mean current in turbulent decomposition ($\mathrm{m\ s^{-1}}$) (Section 2.11)
V_R	tip speed of the rotor blades ($\mathrm{m\ s^{-1}}$) (Section 3.6)
V'	horizontal cross-stream turbulent current component ($\mathrm{m\ s^{-1}}$) (Section 2.11)
W	vertical current in turbulent decomposition ($\mathrm{m\ s^{-1}}$) (Section 2.11)
W_M	vertical mean current in turbulent decomposition ($\mathrm{m\ s^{-1}}$) (Section 2.11)
W'	vertical turbulent current component ($\mathrm{m\ s^{-1}}$) (Section 2.11)
x	Cartesian coordinate in Kepler's first law (m) (Section 1.3)
y	Cartesian coordinate in Kepler's first law (m) (Section 1.3)
z	elevation in, for example, the Bernoulli equation (m) (Section 2.6)
θ	latitude used in Newton's analysis (m) (Section 1.4)
κ	von Karman's coefficient (Section 2.7)
κ_V	van Veen coefficient (Section 2.7)
λ	wavelength of the tide in, for example, the Airy wave (m) (Section 2.8)
μ	dynamic viscosity of water in a Newtonian fluid ($\mathrm{kg\ m^{-1}\ s^{-1}}$) (Section 2.4)
ν	kinematic viscosity, equal to the dynamic viscosity divided by the density in a Newtonian fluid (Section 2.4)
ρ	density of water $= 1026\,\mathrm{kg\ m^{-3}}$
ρ_A	phase difference when $t = 0$ in the harmonic current equation (h) (Section 1.9)
σ	Airy wave radian frequency $= 2p/T$ ($\mathrm{rad\ s^{-1}}$) (Section 2.8)
τ	shear stress in a Newtonian fluid ($\mathrm{N\ m^{-2}}$) (Section 2.4)
T	shaft torque (N m) (Section 3.6)
ϕ	gravitational potential due to the Moon in Newton's equations (Section 1.4)

Part I: Theory

1

History of tidal and turbine science

1.1 Introduction

Renewable energy in general and tidal power in particular are receiving a great deal of attention at present, driven by issues including the security of supply and the availability of hydrocarbon energy sources. Tidal power has traditionally implied the construction of turbines in a barrage for the generation of electricity (cf. Section 3.2 below). This book, however, moves the focus forward to analyse the physics and deployment of tidal stream power devices that are fixed to the seabed and generate electricity directly from the ebbing and flooding of the currents.

The Analysis of Tidal Stream Power   Jack Hardisty
© 2009 John Wiley & Sons, Ltd

This chapter introduces tides and tidal phenomena through consideration of the original work of some of the great innovators. The first sections deal with the early recognition of tidal processes and the rapid progress following the adoption of the heliocentric model for the solar system. Biographical summaries are used to introduce the fundamental work of Isaac Newton and the other luminaries. Attention then turns to the development of turbine science, and progress is reviewed from the earliest water wheels through the turbines of the late eighteenth century to modern turbines and to their application in low-head tidal streams.

The essence of the problem is to be able to understand and to predict the electrical power output from, and the economics of, a tidal stream device throughout periods from a tidal cycle to a year or more. The result is a complex flow that may accelerate from rest to speeds in excess of $3 \, \text{m s}^{-1}$ before peaking, reversing, and accelerating and decelerating in the opposite direction. Within this general, daily, or twice-daily pattern will be the effects of turbulence, which generates peak flows and outputs that are significantly larger than the mean. There are also variations from the seabed to the surface. The flow will peak just below the surface and will decelerate as the seabed is approached. The rate of deceleration depends both upon the overall flow velocity and upon the nature of the seabed. Further complications include the fact that the magnitude of the flows increases and decreases over the lunar cycle, peaking a few days after the new and the full Moons and increasing still further around the equinoxes in March and September. Consideration must then be given to the type of device, to its efficiency in converting flow energy into rotary motion and then into electrical power, and, overall, to the cost of the construction and operational maintenance of the device. This is a fascinating story written alongside 3000 years of cultural, religious, and scientific history.

A very good bibliography of tidal power developments is given by Charlier (2003). This chapter is based, in part, on the books by Defant (1961), Cartwright (1999) and Andrews and Jelley (2007). Cartwright offers a scholarly historical treatise on tidal science, but it is difficult to identify the significant developments within a plethora of players. Here, by sacrificing many of the named contributors (Cartwright lists more than 200, while only 10 are identified here), it is hoped that a clearer derivation of the important principles emerges. These principles are then taken forward in the later chapters. The development of tidal turbines is based upon the first chapter in Round (2004) and the references contained therein. The principles of flow-driven power machines are shown to emerge from the work of Archimedes, Vetruvius, and Poncelet and lead to the modern turbines of Francis, Kaplan, Pelton, and others. Again, these principles are taken forward in later chapters.

Part 1   Tidal science

1.2   Antiquity: Aristotle and Ptolemy

Antiquity forms the earliest period in the traditional division of European history into three 'ages': the classical civilization of Antiquity, the Middle Ages between about AD 500 and AD 1500, and then Modern Times. The early history of tidal science is described by Pugh (1996) who reports that the first evidence of man's interaction with the rise and fall of the tide, implying some understanding of the processes, is the discovery by Indian archaeologists of a tidal dock near Ahmedabad, dating from 2000 BC (Pannikar & Srinivasan, 1971). The earliest known reference to the connection between the tides and the Moon is found in the Samaveda of the Indian Vedic period (2000–1400 BC).

1.2.1 Aristotle and cosmology

The development of tidal science began in Antiquity with the cosmology of Aristotle (Figure 1.1) who observed that 'ebbings and risings of the sea always come around with the Moon and upon certain fixed times'. Aristotle, working in the third century BC, used his books *On the Heavens* and *Physics* to put forward his notion of an ordered universe divided into two distinct parts, the earthly region and the heavens. The earthly region was made up of the four elements: earth, water, air, and fire. Earth was the heaviest, and its natural place was the centre of the cosmos, and for that reason, Aristotle maintained, the Earth was situated at the centre of the cosmos. The heavens, on the other hand, were made up of an entirely different substance, called the aether, and the heavenly bodies were part of spherical shells of aether from the Moon out to Mercury, Venus, the Sun, Mars, Jupiter, Saturn, and the fixed stars. Aristotle argued that the orbits of the heavenly bodies were circular and that they travelled at a constant speed.

Ingenious as this cosmology was, it turned out to be wholly unsatisfactory for astronomy. Heavenly bodies did not move with perfect circular motions: they accelerated, decelerated, and in the cases of the planets even stopped and reversed their motions. Although Aristotle and his contemporaries tried to account for these variations by splitting individual planetary spheres into components, these constructions were very complex and, ultimately, doomed to failure. Furthermore, no matter how complex a system of spheres for an individual planet became, these spheres were still

Figure 1.1 Aristotle (384–322 BC).

centred on the Earth. The distance of a planet from the Earth could therefore not be varied in this system, but planets varied in brightness. Since variations in intrinsic brightness were ruled out, and since spheres did not allow for a variation in planetary distances from the Earth, variations in brightness could not be accounted for in this system.

Other developments in tidal science at this time included those by Pytheas who travelled through the Strait of Gibraltar to the British Isles and reported the half-monthly variations in the range of the Atlantic Ocean tides, and that the greatest ranges (Spring tides) occurred near the new and the full Moons.

Many other aspects of the relationship between tides and the Moon are noted in Pliny the Elder's (AD 23–79) *Natural History*. Pliny described how the maximum tidal ranges occur a few days after the new or full Moon, and how the tides at the equinoxes in March and September have a larger range than those at the summer solstice in June and winter solstice in December.

1.2.2   Ptolemy's geometrical solar system

Mathematicians who wished to create geometrical models of the solar system in order to account for the actual motions of heavenly bodies began using different constructions within a century of Aristotle's death. Although these violated Aristotle's physical and cosmological principles somewhat, they were ultimately successful in accounting for the motions of heavenly bodies. We see the culmination of these efforts in the work of Claudius Ptolemy (Figure 1.2). In his great astronomical work *Almagest*, Ptolemy presented a complete system of mathematical constructions that accounted successfully for the observed motion of each heavenly body. Ptolemy used three basic constructions, the eccentric, the epicycle, and the equant, to describe the movements of the planets, the Sun, and the Moon. With such combinations of constructions, Ptolemy was able to account for the motions of heavenly bodies within the standards of observational accuracy of his day.

Figure 1.2   Claudius Ptolemy (AD 90–168).

Early explanations for the tides were curious; Aristotle is credited with the law that no animal dies except when the water is ebbing. This idea survived into popular culture. For example, as recently as 1595 in the North of England, the phase of the tide was recorded at the time of each person's death. Eastern cultures held the belief that the water was the blood of the Earth and that the tides were caused by the Earth breathing.

1.3 Middle Ages: Copernicus to Galileo

The Middle Ages was a period of great cultural, political, and economic change in Europe and is typically dated from around AD 500 to approximately AD 1500. During the early Middle Ages, for example, the Venerable Bede (673–735) was familiar with the tides along the coast of Northumbria in England, and was able to calculate the tides using the 19 year lunar cycle. By the early ninth century, tide tables and diagrams showing how Neap and Spring tides alternate were appearing in several manuscripts.

1.3.1 Copernicus's heliocentric solar system

Although Aristotelian cosmology and Ptolemaic astronomy were still dominant, it was the fundamental change brought about by Copernicus's heliocentric view that removed the barriers to progress and opened the way for the advancement of tidal science by Newton's deterministic analysis in the seventeenth century.

The Polish astronomer Copernicus (Figure 1.3) proposed that the planets have the Sun as the fixed point to which their motions are to be referred and that the Earth is a planet that, besides orbiting the Sun annually, also turns once daily on its own axis. He also recognized that the very slow long-term changes in the direction of this axis account for the precession of the equinoxes. This of the heavens is

Figure 1.3 Nicolaus Copernicus (1473–1543).

usually called the heliocentric, or 'Sun-centred', system – derived from the Greek *helios*, meaning 'Sun'. Copernicus wrote about these ideas in a manuscript called the *Commentariolus* ('Little Commentary') during the period 1508–1514. However, the work that contains the final version of his theory, *De revolutionibus orbium coelestium libri vi* ('Six Books Concerning the Revolutions of the Heavenly Orbs'), did not appear in print until 1543, the year of his death.

1.3.2   Tycho Brahe's observations

Tyge (latinized as Tycho) Brahe (Figure 1.4) was born in Skane, Sweden (The Galileo Project, 2007). He attended the Universities of Copenhagen and Leipzig, and then travelled through Germany, studying at Wittenberg, Rostock, and Basel. His interest in astronomy was aroused during this period, and he bought several astronomical instruments. Tycho Brahe lost part of his nose in a duel with another student, in Wittenberg in 1566, and for the rest of his life he wore a metal insert to cover the scar. He returned to Denmark in 1570.

Brahe accepted an offer from King Frederick II in the 1570s to fund an observatory. He was given the little island of Hven in the Sont near Copenhagen, and there he built Uraniburg, which became the finest observatory in Europe. Brahe designed and built new instruments, calibrated them, and instituted nightly observations. He also ran his own printing press. His observations were not published during his lifetime, but he employed Johannes Kepler as an assistant to calculate planetary orbits from the data. Brahe did not entirely abandon the Copernican, Earth-centred approach because, he argued, if the Earth were not at the centre of the universe, physics, as it was then known, was utterly undermined. Instead, he developed a system that combined the best of both worlds. He kept the Earth in the centre of the universe, so that he could retain Aristotelian physics. The Moon and Sun revolved

Figure 1.4   Tycho Brahe (1546–1601).

about the Earth, and the shell of the fixed stars was centred on the Earth. But Mercury, Venus, Mars, Jupiter, and Saturn revolved about the Sun. This Tychonic world system became popular early in the seventeenth century among those who felt forced to reject the Ptolemaic arrangement of the planets (in which the Earth was the centre of all motions) but who, for various reasons, could not accept the Copernican alternative.

1.3.3   Kepler's laws of planetary motion and the 'sphere of influence of the Moon' tidal theory

Brahe's student Johannes Kepler (Figure 1.5) used simple mathematics to describe how planets move. Kepler was able to use Tycho Brahe's data to develop three empirical laws that described the movement of the planets and that supported the heliocentric Copernican system (Drennon, 2007):

- **Kepler's first law** states that the orbit of a planet about the Sun is an ellipse with the Sun's centre of mass at one focus. Thus, the orbit of a planet (Figure 1.6) is given by

$$\frac{x^2}{a_K^2} + \frac{y^2}{b_K^2} = 1 \tag{1.1}$$

where x and y are the Cartesian coordinates, and a_K and b_K are the lengths of the semi-minor and semi-major axes respectively. (Note that all symbols are defined

Figure 1.5   Johannes Kepler (1571–1630).

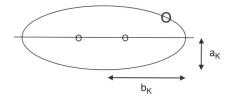

Figure 1.6   Johannes Kepler's first law of planetary motion.

in the Notation at the beginning of the book and will rarely be defined henceforth in the text.)

- **Kepler's second law** states that a line joining a planet and the Sun sweeps out equal areas in equal intervals of time.

- **Kepler's third law** states that the squares of the periods of the planets are proportional to the cubes of their semi-major axes:

$$\frac{T_1^2}{T_2^2} = \frac{R_1^3}{R_2^3} \tag{1.2}$$

where T_1 and T_2 are the sidereal periods of orbit, and R_1 and R_2 are the lengths of the semi-major axes of planets 1 and 2 respectively. Kepler also proposed a theory for the tides when he wrote, in 1609, that 'The sphere of influence of the attraction which is in the Moon extends as far as the Earth, and incites the waters up from the torrid zone...', thus invoking the force that Isaac Newton would later call gravity.

1.3.4 Galileo's differential motion tidal theory

Although Galileo Galilei (Figure 1.7) probably developed his explanations for tidal phenomenon within the new heliocentric Copernican system in the closing years of

Figure 1.7 Galileo Galilei (1564–1642).

the sixteenth century, they were not formally published until 1616 in his *Treatise on the Tides* (Tyson, 2007). The idea is said to have occurred to him while travelling on a barge ferrying fresh water to Venice, close to his home in Padua. He noticed that, whenever the barge's speed or direction altered, the fresh water inside sloshed around accordingly. If the vessel suddenly ground to a halt on a sandbar, for instance, the water pushed up towards the bow then bounced back towards the stern. Galileo proposed that the Earth's dual motion – its daily rotation around its axis and its annual rotation around the Sun – might have the same effect on oceans and other bodies of water as the barge had on its freshwater cargo. The key, as Galileo saw it, was that different parts of the planet moved at different speeds depending on the time of day, as if the Earth were a barge that accelerated and decelerated and periodically changed direction.

1.3.5   Descartes' universal matter tidal theory

By the mid-seventeenth century. three different theories for the origins of the tides were being seriously considered:

1. Kepler was one of the originators of the idea that the Moon exerted a gravitational attraction on the water of the ocean, drawing it towards the place where it was overhead. This attraction was balanced by the Earth's attraction on the waters, for 'If the Earth should cease to attract its waters, all marine waters would be elevated and would flow into the body of the Moon'.

2. Galileo proposed that the rotations of the Earth produced motions of the sea, which were modified by the shape of the seabed to give the tides.

3. The French mathematician, philosopher, and scientist Rene Descartes (Figure 1.8) introduced the third theory in the early part of the seventeenth century. Descartes argued that space was full of invisible matter. As the Moon

Figure 1.8   Renee Descartes (1596–1650).

travelled around the Earth, it compressed the matter in a way that transmitted pressure to the sea, hence forming the tides.

Although there was merit in each of these approaches, it was Isaac Newton's introduction of the concept of force in a heliocentric solar system that was finally to resolve the differences between them.

1.4   Isaac Newton and the equilibrium theory

The newly established Royal Society of London was a rich and fertile environment for the development of tidal science in the middle years of the seventeenth century. Association d'Oceanographique Physique (1955) lists a number of papers on tidal phenomena in the first volumes of the *Transactions* between the years 1665 and 1668, including Moray (1665, 1666a and b), Wallis (1666), Colepresse (1668), Norwood (1668), Philips (1668), Stafford (1668), and Wallis (1668).

A major step forward in the understanding of the processes by which tides are generated was made by the British mathematician, Isaac Newton (Figure 1.9). Newton was able to apply his formulation of the law of gravitational attraction to show why there were two tides for each lunar transit, why the Spring to Neap cycle occurred, and why daily tides were a maximum when the Moon was furthest from the plane of the equator (Pugh, 1996).

Newton's seminal work, which transformed the worlds of physics and mathematics, begins with the concept that there is an inverse and linear relationship between the acceleration and the mass of a body. This is captured by the simple proportionality

$$a \alpha \frac{1}{m} \tag{1.3}$$

Figure 1.9   Isaac Newton (1643–1727).

The coefficient of proportionality was called the 'force', so that the relationship is captured in Newton's first two laws of motion:

- **Newton's first law** states that every object in a state of uniform motion tends to remain in that state of motion unless an external force is applied to it.

- **Newton's second law** states that the relationship between an object's mass, its acceleration, and the applied force is

$$F = ma \qquad (1.4)$$

The unit of force is, of course, the newton. Newton's *Principia* was published in 1687 and proposed and then used the problem of tides to prove that the gravitational force between two masses depends upon the product of the masses and the inverse square of their separation:

$$F = \frac{GMm}{R^2} \qquad (1.5)$$

where G is the universal gravitational constant which is now known to have a value of $6.67300 \times 10^{-11}\,\mathrm{m^3\,kg^{-1}\,s^{-2}}$. Equations (1.4) and (1.5) are combined to derive Newton's equilibrium theory of the tides (e.g. Andrews & Jelley, 2007). We now know that Kepler was right and that the main cause of the tides is the gravitational effect of the Moon. The effect of the Sun is about half that of the Moon but increases or decreases the size of the lunar tide according to the positions of the Moon and the Sun relative to the Earth. Galileo was wrong; the daily rotation of the Earth about its own axis only affects the location of the high tides. Galileo was right, however, in his advocacy of the heliocentric solar system. We initially ignore the effect of the Sun in the following derivation.

For simplicity, we consider that the Earth is covered by water, as shown schematically in Figure 1.10. Consider a unit mass of water situated at some point P. The gravitational potential due to the Moon is then given by

$$\phi = -\frac{GM_{\mathrm{M}}}{s} \qquad (1.6)$$

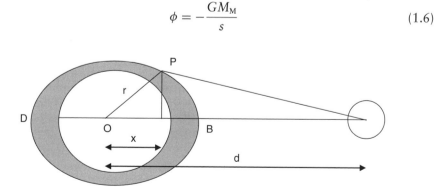

Figure 1.10 Tidal effect of the Moon (not to scale).

where s is the distance between the centre of mass of the Moon and point P. For $d \gg r$ we can expand $1/s$ as follows by applying Pythagoras' theorem to the triangle OPN in the figure:

$$\frac{1}{s} = \frac{1}{[d^2 + r^2 - 2dr\cos\theta]^{1/2}} = \frac{1}{d}\left[1 + \left(-\frac{2r}{d}\cos\theta + \frac{r^2}{d^2}\right)\right]^{1/2}$$

$$= \frac{1}{d}\left[1 + \frac{r}{d}\cos\theta + \frac{r^2}{d^2}\left(\frac{3}{2}\cos^2\theta - \frac{1}{2}\right) + L\right] \tag{1.7}$$

The first term in the expansion does not yield a force and can be ignored. The second term corresponds to a constant force Gm/d^2 directed towards N which acts on the Earth as a whole and is balanced by the centrifugal force due to the rotation of the Earth–Moon system. The third term describes the variation of the Moon's potential around the Earth. The surface profile of the water in this simplified system is an equipotential surface due to the combined effects of the Moon and the Earth. The potential of a unit mass of water owing to the Earth's gravitation is gh and $g = GM_E/r^2$. Hence, the height of the tide at θ is given by

$$gh(\theta) - \frac{Gmr^2}{d^3}\left(\frac{3}{2}\cos^2\theta - \frac{1}{2}\right) = 0 \tag{1.8}$$

or

$$h(\theta) = h_{max}\left(\frac{3}{2}\cos^2\theta - \frac{1}{2}\right) \tag{1.9}$$

where

$$h_{max} = \frac{mr^4}{Md^3} \tag{1.10}$$

The maximum height of the tide, h_{max}, occurs at points P and D, where $\theta = 0$ and $\theta = \pi$. Putting $m/M = 0.0123$, $d = 384\,000$ km, and $r = 6378$ km, we obtain $h_{max} = 0.36$ m, which is roughly in accord with the observed mean tidal height. Equation (1.9) was incorporated into a software routine (which is available from this book's online website) and produces the global variation in the tidal heights shown in Figure 1.11, with the two maxima corresponding to high water on the near side and far side of the Earth from the Moon.

Newton's equilibrium theory finally reconciled science and religion and, through the introduction of the new physics of force and response, explained, in a quasi-static sense, the behaviour of the waters of the world's oceans and the ebbing and flowing of the tides. Although the explanation was correct, it could not be applied to particular sites without the advantage of long-term datasets, as described in the following section.

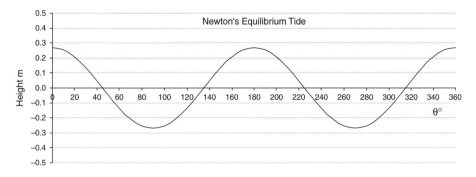

Figure 1.11 The global variation in the height of the lunar ride (from equation (1.9)), utilizing the model from this book's online website.

1.5 Measurement after Moray

The development of observational techniques has to be viewed as a necessary progression that happened alongside the establishment of the theoretical framework. The problem was, of course, that, although Newton's gravitational theory provided the fundamental explanation for tidal phenomena, it was not possible to apply the theory on a local basis to predict the height and timing of high water and low water at a specific site because of the complications of bathymetry and coastal form. Instead, long continuous records of water depth (as opposed to approximate measurements at points in time) were to lead, eventually, to the identifications of the tidal components and to the harmonic analysis that powerfully provided accurate local predictions.

The early Royal Society of London was strongly inclined to experimentation, and was the first institution to encourage careful series of tidal measurements as distinct from the collection of casual reports from mariners. A formal set of recommendations was assigned to Sir Robert Moray who, in 1666, described a stilling well with a small connection to the sea and within which was a float connected to a pen recorder (Moray, 1666a). It was more than 150 years later that the first continuous sea level record to cover a Spring–Neap cycle was taken (Figure 1.12). The measurements were obtained by a civil engineer named Mitchell and appear to have been based upon a design suggested to the Royal Society by Henry Palmer (1831) which had its antecedents in Moray's work. These gauges are therefore known as Palmer–Moray devices.

At the same time as these temporal variations in tidal heights were being determined, workers such as the Reverend William Whewell and J.W. Lubbock were involved in defining spatial variations in the range and progress of the tides. Admiral Sir Francis Beaufort (1774–1857) (Figure 1.13), hydrographer to the British Royal Navy, assisted in their work. As well as providing staff to assist Lubbock's efforts, Beaufort arranged for tides to be observed at over 100 British coastguard stations for 2 weeks in June 1884. The following year, another observational exercise was achieved, covering 101 ports in seven European countries, 28 in America from the Mississippi to Nova Scotia, and 537 ports in the British Isles including Ireland.

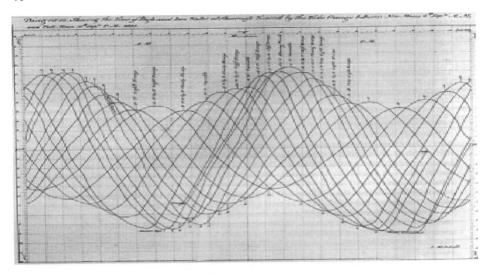

Figure 1.12 The first known record of a continuous Spring–Neap cycle taken at Sheerness Dockyard on the River Thames in England from 6 to 21 September 1831. From Cartwright (1999).

Figure 1.13 Admiral Sir Francis Beaufort (1774–1857).

After more than 100 years of such observations around the world, Figure 1.14 has been constructed to show the global distribution of tidal range and the cotidal lines. Much of the tidal motion has the character of rotary waves. In the south and equatorial Atlantic Ocean the tide mainly takes the form of north–south oscillation on east–west lines. This complex reality may usefully be compared with the simple equilibrium concept which pictures the tide as a sinusoidal wave progressing around the Earth in the easterly direction.

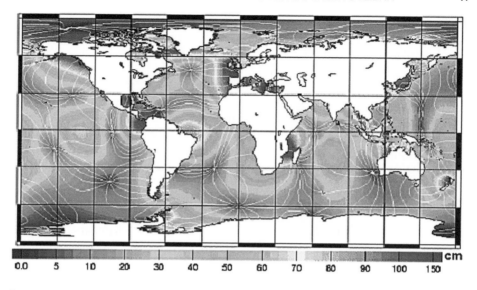

Figure 1.14 Global distribution of tidal range and cotidal lines (LEGOS, 2008). A URL link to the full colour version of this image is available on the book's website.

1.6 Eighteenth and nineteenth centuries: Laplace to Kelvin

1.6.1 Laplace tidal equations

Pierre Simon Laplace (Figure 1.15) was born at Beaumont-en-Auge in Normandy (Rouse, 1908). He produced much of his original work in astronomy during 1771–1787, commencing with a 1773 memoir showing that the planetary motions were stable. This was followed by several papers on points in the integral calculus, finite differences, differential equations, and astronomy. During the years 1784–1787 he

Figure 1.15 Pierre Simon de Laplace (1749–1827).

produced some memoirs that were reprinted in the third volume of the *Méchanique Céleste*, in which he completely determined the attraction of a spheroid on a particle outside it. This is memorable for the introduction of spherical harmonics which are now known as Laplace's coefficients.

Cartwright (1999) explains that Laplace's major theoretical advance was the formulation of a system of three linear, partial differential equations for the horizontal components of ocean velocity relative to the Earth, and the vertical displacement of the ocean surface. Known to tidal scientists as the Laplace tidal equations (LTEs), the equations remain the basis of tidal computation to this day. Laplace also showed that the equations were dynamically stable for realistic values of the physical constants, and therefore the frequencies of any solutions had to coincide with frequencies present in the forcing term. There were many such frequencies, due to various astronomical periodicities. From the fact that the trigonometric expansion contained terms involving the cosines of t and $2t$, as well as time-independent arguments, Laplace concluded that there were at least three 'species' of tidal waves. The first species, corresponding to terms independent of time, were long-period tides – typically of low amplitude – which would not be apparent to the casual dockside observer but which careful measurements had already begun to detect. The second, corresponding to terms involving t, were the main cause of the long-noted differences between successive high- or low-water heights. The third, corresponding to terms involving $2t$, caused the familiar semi-diurnal tides.

1.6.2 Kelvin and the harmonic species

From about 1833, the British Admiralty first began to publish tide tables with the times and heights of the tidal extrema, high water and low water, based upon the 'synthetic method' of J.W. Lubbock which involved very long and laborious cross-correlation of the extrema with lunar constants and timings. The first official British Admiralty predictions covered the four English ports of Plymouth, Portsmouth, Sheerness, and London Bridge.

The subject was taken forward by William Thompson, the 1st Baron Kelvin (Figure 1.16). Kelvin appreciated that the new continuous tidal records would permit the analysis of the whole tidal profile and more accurate and easier tidal predictions. Kelvin was a mathematical physicist, engineer, and outstanding leader in the physical sciences of the nineteenth century. He did important work in the mathematical analysis of electricity and thermodynamics and in unifying the emerging discipline of physics in its modern form. He is widely known for developing the Kelvin scale of absolute temperature measurement. The title Baron Kelvin was given in honour of his achievements, and named after the River Kelvin, which flowed past his university in Glasgow, Scotland.

Kelvin devised the method of reduction of tides by harmonic analysis about the year 1867. Harmonic analysis is based on the principle that any periodic motion or oscillation can always be resolved into the sum of a series of simple harmonic motions (Phillips, 2007). The principle is said to have been discovered by Eudoxas in 356 BC, when he explained the apparently irregular motions of the planets by combinations of uniform circular motions. In the early part of the nineteenth century, Laplace recognized the existence of partial tides that might be expressed by

Figure 1.16 Lord Kelvin (1824–1907).

Table 1.1 The astronomical time constants which give the fundamental frequencies for gravitational tides (MSD = mean solar day)

Name			Period
Mean lunar day	T_1	The time of the rotation of the Earth with respect to the Moon, or the interval between two upper transits of the Moon over the meridian of a place	1.035050 MSD
Tropical month	s	The average period of the revolution of the Moon around the Earth with respect to the vernal equinox	27.321582 MSD
Tropical year	h	The average period of the revolution of the Earth around the Sun with respect to the vernal equinox	365.2422 MSD
Rotation of Moon's perigee	p	Change in mean longitude of the lunar perigee	8.85 years
Rotation of Moon's node	N	Change in mean longitude of the Moon's node	18.61 years
Rotation of Earth's perigee	p_1	Change in mean longitude of the solar perigee	20 900 years

the cosine of an angle increasing uniformly with time, and also applied the essential principles of harmonic analysis to the reduction of high and low waters. The earliest analysis by Kelvin made use of combinations of the six main periods shown in Table 1.1 to integrate the effect of the 12 main harmonics shown in Table 1.2.

1.6.3 Fennel and Harris

The American mathematicians who have had an important part in the development of this subject include William Ferrel and Rollin Harris, both of whom were associated with the US Coast and Geodetic Survey. *The Tidal Researches*, by Ferrel, was

Table 1.2 Selected main constituents of the tide and their periods from Table 1.1 and Groves-Kirkby *et al.* (2006)

	Name		Period
S_{sa}	Solar semi-annual constituent	Accounts for the non-uniform changes in the Sun's declination and distance. Mostly reflects yearly meteorological variations influencing the sea level	$2h$
M_m	Lunar monthly constituent	The effect of irregularities in the Moon's rate of change of distance and speed in orbit	$s - p$
M_f	Lunar fortnightly constituent	The effect of departure from a sinusoidal declinational motion	$2s$
O_1	Lunar diurnal constituent	See K_1	$T_1 - 2s + p$
Q_1	Larger lunar elliptic diurnal constituent	Modulates the amplitude and frequency of the declinational O_1	$T_1 - s$
K_1	Lunisolar diurnal constituent	Expresses, with O_1, the effect of the Moon's declination. They account for diurnal inequality and, at extremes, diurnal tides. With P_1, it expresses the effect of the Sun's declination	$T_1 + s$
P_1	Solar diurnal constituent	See K_1	$T_1 + s - 2h$
L_2	Smaller lunar elliptic semi-diurnal constituent	Modulates, with N_2, the amplitude and frequency of M_2 for the effect of variation in the Moon's orbital speed owing to its elliptical orbit	$2T_1 + s - p$
N_2	Larger lunar elliptic semi-diurnal constituent	See L_2	$2T_1 - s + p$
M_2	Principal lunar semi-diurnal constituent	The rotation of the Earth with respect to the Moon	$2T_1$
S_2	Principal solar semi-diurnal constituent	The rotation of the Earth with respect to the Sun	$2T_1 + 2s + 2h$
K_2	Lunisolar semi-diurnal constituent	Modulates the amplitude and frequency of M_2 and S_2 for the declinational effect of the Moon and Sun respectively.	$2T_1 + 2s$

published in 1874, and additional articles on harmonic analysis by the same author appeared from time to time in the annual reports of the Superintendent of the Coast and Geodetic Survey. The best known work of Harris is his *Manual of Tides*, which was published in several parts as appendices to the annual reports of the Superintendent of the Coast and Geodetic Survey. The subject of harmonic analysis was treated principally in Part II of the *Manual* which appeared in 1897.

1.7 Tide-predicting machines

Lord Kelvin also designed the first tide-predicting machine, which was built in 1873. It was based on a suggestion by Beauchamp Towers for summing several trigonometric functions with independent periods and under the auspices of the British

Association for the Advancement of Science (Kelvin, 1911; FWERI, 2007; NOAA, 2007a, 2007b). The origins of the idea are described by AMS (2007) as follows:

> In 1872 Kelvin was preoccupied with the tidal prediction, and in particular with the problem of summing a large number of harmonic motions with irrationally related frequencies. Here is how the solution came to him, in his own words (*Mathematical and Physical Papers, Vol. VI.* Cambridge, 1911, p. 286):
>
>> On his way to attend the British Association in 1872, with Mr. Tower for a fellow-passenger, the Author was deeply engaged in trying to find a practical solution for the problem. Having shown his plans and attempts to Mr. Tower, whose great inventiveness is well known, Mr. Tower suggested: '*Why not use Wheatstone's plan of the chain passing around a number of pulleys, as in his alphabetic telegraph instrument?*' This proved the very thing wanted. The plan was completed on the spot; with a fine steel hair-spring, or wire, instead of the chain which was obviously too frictional for the tide predictor. Everything but the precise mode of combining the several simple harmonic motions had, in fact, been settled long before. At the Brighton meeting … the Author described minutely the tide-predicting machine thus completed in idea, and obtained the sanction of the Tidal Committee to spend part of the funds then granted to it on the construction of a mechanism to realise the design for tidal investigation by the British Association. Before the end of the meeting he wrote from Brighton to Mr. White at Glasgow, ordering the construction of a model to help in the designing of the finished mechanism for the projected machine.

The device was an integrating machine designed to compute the height of the tide in accordance with

$$h = H_0 + \sum \left\{ fH \cos \left[a_A t + (V_0 + u) - K \right] \right\} \tag{1.11}$$

Tide-predicting machines are simple devices that essentially use mechanical means to represent the trigonometric functions (Figure 1.17). Kelvin's first machine provided for the summation of 10 of the principal constituents M_2, S_2, N_2, K_1, O_1, K_2, L_2, P_1, M_4, and MS_4, and the resulting predicted heights were registered by a pen trace. The machine is described in Part I of Thomson and Tait's *Natural Philosophy*, 1879 edition. Kelvin's second machine was a smaller model completed in about 1880 and was in turn succeeded by a third in about 1883 and a 'Fourth British Tide Predictor' in about 1910.

The first tide-predicting machine used in the United States was designed by William Ferrel, of the US Coast and Geodetic Survey, and completed in 1882. Ferrel's machine summed 19 harmonic tidal constituents and gave direct readings of the predicted times and heights of the high and low waters. Predictions were made on this machine from 1885 until 1991. Ferrel's machine did not trace an output curve, and the times and heights of the high and low waters were indicated directly by scales on the machine. The Coast and Geodetic Survey in the United States constructed a second machine that provided continuous, graphical outputs of the predictions in 1910. The machine was about 11 ft long, 2 ft wide, and 6 ft high and weighed approximately 2500 lb. NOAA (2007a, 2007b) also reports tide-predicting machines that were built in Brazil and Germany.

Figure 1.17 The first and second Kelvin tidal machines.

Figure 1.18 (Left) Arthur Doodson (1890–1968) and (right) Arthur Doodson's daughter-in-law, Valerie, operates the Doodson–Légé tidal machine at the Bidston Observatory.

In the United Kingdom, Kelvin's machine continued to be used and was more permanently installed at the Bidston Tidal Observatory on Merseyside until the Doodson–Légé tide-predicting machine was designed by one of the Observatory's directors, A.T. Doodson, and built by the Liverpool firm Légé during 1948/1949. The machine resolved up to 42 constituents and was in daily use from then until the early 1960s, when it was superseded by the electronic computer (Figure 1.18).

1.8   Tidal currents

The measurement of a ship's speeds through the water, and conversely the measurement of tidal currents past a moored ship, probably date back into antiquity.

Tidal currents are routinely reported in units of knots. The definition of the knot (symbolized by kn or kt) is based on the internationally agreed length of the nautical mile, since a speed of 1 knot is equivalent to one nautical mile per hour and exactly equal to 1.852 km. The international definition was adopted by the USA in 1954 (which previously used the US nautical mile of 1853.248 m). The international definition was adopted by the UK in 1970. The knot is approximately equivalent to 101.268591 feet per minute, 1.687810 feet per second, 1.150779 miles (statute) per hour (mph), or 0.5144444 recurring metres per second. For convenience, equivalence to 0.51 m s^{-1} is usually used.

Until the mid-nineteenth century, vessel speed at sea (or current speed past a moored vessel) was measured using a chip log (Art of Navigation, 2007). This device consisted of a wooden panel, weighted on one edge to float upright, attached by line to a reel. The chip log was 'cast' over the stern of the moving vessel and the line allowed to pay out. Knots placed at a distance of 47 ft 3 in (14.4018 m) passed through a sailor's fingers, while another sailor used a 30 s sandglass to time the operation. This method gives a value for the knot of 20.25 in s^{-1}, or $1.85166 \text{ km h}^{-1}$. The difference from the modern definition is less than 0.02 %.

The Proudman Oceanographic Laboratory (2007a, 2007b) reports on the more recent and instrumented history of tidal-current metering. Figure 1.19 (left) shows Arthur Doodson with an early current meter. The water movements turn the large fan at the front, and the vane at the back orients the meter with the direction of the current flow.

Figure 1.19 (right) shows a current meter record which was probably written by a mechanical stylus onto a smoked glass slide. It is about 2 min long. The average current is about 0.7 kt (0.36 m s^{-1}): this will be due to a steady tidal current. The average current is modulated by a signal of about 15 cycles per minute: this will be caused by small surface waves with a 4 s period.

NOAA (2007a, 2007b) notes that, since the tidal-current velocities in any locality may be expressed by the sum of a series of harmonic terms involving the same periodic constituents as those found in the tidal elevations, the tide-predicting machine may also be used for tidal-current prediction. The harmonic constants for the prediction of current velocities are derived from current observations by an analysis

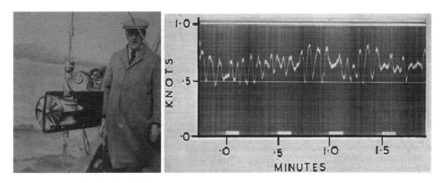

Figure 1.19 (Left) Arthur Doodson with an early current meter and (right) an early current meter record.

similar to that used in obtaining the harmonic constants from tidal height observations. For the currents, however, consideration must be given to the direction
of flow.

The machine can be used for the prediction of reversing currents in which the
direction of the flood current is taken as positive and the ebb current as negative. Rotary currents may be predicted by taking the north and east components
separately, but the labour of obtaining the resultant velocities and directions from
these components would be very great without a machine especially designed for
the purpose.

Kelvin's equation is then rewritten as the harmonic current equation:

$$U(t) = U_0 + \sum \left[U_A \cos \left(\frac{2\pi t}{T_A} + \rho_A \right) \right] \qquad (1.12)$$

Equation (1.12) is, in effect, the result of all of the work stretching for more than two
millennia and described in Part I of this chapter. We have seen how a rudimentary
knowledge of astronomy developed by the Ancient Greeks has developed into a very
detailed understanding of planetary motion. Newton has then used the gravitational
attraction of these orbiting masses to explain the equilibrium theory of the tides, and
Laplace and Kelvin have then developed the harmonic analysis of tidal datasets to
explain and to predict the vertical and now the horizontal motions of tidal waters.
In the following sections, we examine the development of turbine science to utilize
the water currents and develop the analysis of tidal stream power.

Part 2   Turbine science

1.9   Antiquity: the Romans and Chinese

From classical times (Hansen, 2007) there have existed three general types of water
wheel (Figure 1.20). The Norse wheel has vanes protruding from a wooden rotor
which is turned by a jet of water. The undershot wheel is rotated by the impulse

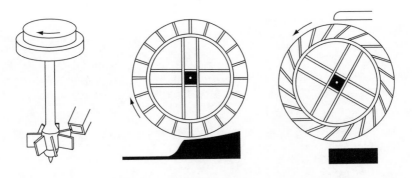

Figure 1.20 Water wheels. Norse wheels (left) turn millstones directly, undershot
wheels (centre) require gears, and overshot wheels (right) also require an elevated
stream (after Hansen, 2007).

Figure 1.21 Marcus Vitruvius Pollio (ca 80 to ca 25 BC).

of a water current. The overshot wheel receives water from above and utilizes both the potential energy of the mass of the water and the kinetic energy of the current. It seems probable (Ramage, 2004) that the earliest water device was an undershot wheel known as the noria and used to raise water for irrigation.

The first description of a water wheel is from Vitruvius (Figure 1.21), an engineer of the Augustan Age (31 BC to AD 14) who composed a 10-volume treatise on all aspects of Roman engineering. Vitruvius described the workings of an undershot wheel. One innovation occurred when Rome was under siege in AD 537 and the besieging Goths shut off the aqueducts that drove the city's mills. Belisarius, the Byzantine general defending the city, ordered floating mills to be installed close to the Tiber bridges, the piers of which constricted and accelerated the current. Two rows of boats were anchored, with water wheels suspended between them. The arrangement worked so well that cities all over Europe were soon copying it.

Water power was also an important source of energy in ancient Chinese civilizations. A contemporary text reports that, in AD 31, the engineer Tu Shih invented a water-powered reciprocator for the casting of agricultural implements. Smelters and casters were 'instructed to use the rushing of water to operate their billows'. Water power was also applied at an early date to the grinding of grain. Large mills appeared in China in about the second century BC and at about the same time in Europe. However, in China the water wheel was a critical power supply, while, for hundreds of years, Europe relied heavily on slave- and donkey-powered mills.

1.10 Middle Ages: the Syrians and Agricola

The fall of the western part of the Roman Empire in about AD 400 was followed by a transitional period of about 1000 years. Advances in hydraulics that had occurred under such notables as Archimedes effectively ceased as Europe became fragmented into small states. In the Arab world, however, knowledge of water supply using combinations of water wheels and Archimedean screws flourished. The water wheels that had been originally introduced into Europe by the Romans were improved during this period, and the Middle East developed many ingenious combinations of screws and undershot and overshot water wheels (Sarton, 1931). In contrast, in the Western World, the structures that had been built by the Romans were allowed to deteriorate.

There was, however, a gradual spread in the use of traditional water wheels. Areas in northern and western Europe came under cultivation. Grain was an important crop, and most of it was ground by water mills (e.g. ETMTL, 2007). Historic records provide useful insights. The Doomsday Book, a survey prepared in England in AD 1086, lists 5624 water mills, whereas, only a century earlier, fewer than 100 mills were counted. French records tell a similar story. In the Aube district, 14 mills operated in the eleventh century, 60 in the twelfth, and nearly 200 in the thirteenth century. In Picardy, 40 mills in 1080 grew to 245 by 1175. Boat mills, moored under the bridges of medieval Paris and other cities, began in the twelfth century to be replaced by structures permanently joined to bridges. A good picture of metallurgy and water wheels can be obtained from *De Re Metallica*, by Georgius Agricola, published in 1556. An excellent translation of this work was prepared by Herbert Hoover (a mining engineer and future President) and his wife (the first woman geologist to graduate from Stanford University). Agricola was one of the first to record mining and metallurgical practices, and in so doing has left us with impressive images of water wheel technology.

1.11   Eighteenth and nineteenth centuries: Smeaton to Kaplan

1.11.1   John Smeaton

Major improvements were made in water wheel design from the 1700s onwards. John Smeaton (Figure 1.22) made a series of experiments on model water wheels in 1752 and was elected Fellow of the Royal Society in 1753. In 1759 he published a paper on water wheels and windmills (Smeaton, 1759), for which he received the Copley Medal of the Royal Society. Among his results, he noted that overshot wheels were twice as efficient as undershot wheels and that the impact of streams of water on a flat plate dissipated a large amount of energy. He was a member of the Royal Society Club, an occasional guest at meetings of the Lunar Society, and a charter member of the first professional engineering society, the Society of Civil Engineers. The latter became known as the Smeatonian Society after his death.

1.11.2   Benoît Fourneyron

In 1832 the young French engineer Benoît Fourneyron (Figure 1.23) patented the first water turbine (from the Latin *turbo*, meaning something that spins). Fourneyron's turbine incorporated guide vanes to direct the water towards the centre across the curved faces of the fixed vanes so that it travelled almost parallel to the curve of the runner blades as it reached them. The water was deflected as it crossed the

Figure 1.22   John Smeaton (1724–1792).

Figure 1.23   Benoît Fourneyron (1802–1867).

runner blades, exerting a rotational force, and then, having given up its energy, it fell away into the outflow. Tests showed that Fourneyron's turbine converted as much as 80 % of the energy in the water into useful mechanical output.

1.11.3   James Bicheno Francis

Soon after Fourneyron's turbine developments, the British–American engineer James Bicheno Francis (Figure 1.24) invented the Francis inward flow radial turbine (combining radial and axial flow) for low-pressure installations. He had joined the Locks and Canal Company of Lowell, Massachusetts, shortly after emigrating to the United States at the age of 18. He became Chief Engineer in 1837 and remained

Figure 1.24   James Bicheno Francis (1815–1892).

with the Company for his entire career. The Company owned and operated Lowell's canal system, providing water power for the local textile mills.

1.11.4   Lester Allen Pelton

During the nineteenth century, a number of impulse turbines were also invented. Lester Pelton (Figure 1.25) developed the first water wheel to take advantage of the kinetic energy of water rather than the weight or pressure of a stream. The speed and efficiency of Pelton's wheel made it ideal for generating electricity. Pelton designed a wheel with split buckets that harnessed the kinetic energy of a small volume of water flowing at high speed. Properly adjusted, Pelton's wheel could achieve efficiencies in

Figure 1.25   Lester Allen Pelton (1829–1908).

Figure 1.26 Victor Kaplan (1876–1934).

excess of 90 %. Pelton's wheel permitted low-cost hydroelectric power to replace expensive steam engines in mining operations in the western United States where streams rarely flowed at high enough volumes to turn traditional water wheels.

1.11.5   Victor Kaplan

Victor Kaplan (Figure 1.26) created the Kaplan turbine, a propeller-type machine, in 1913. It was an evolution of the Francis turbine but revolutionized the ability to develop low-head hydro sites.

1.12   Modern turbines

A modern water turbine is a device that converts the energy in a stream of fluid into mechanical energy by passing the stream through a system of fixed and moving blades and causing the latter to rotate (Electrical Engineering Tutorials, 2008).

Turbines may be classified according to the direction of the water flow into four groups (cf. Chapter 3):

1. **Tangential** or **peripheral turbines** have flow along the tangential direction, such as the Pelton wheel.

2. **Inward radial-flow turbines** have flow along the radius, such as the crossflow turbine.

3. **Mixed flow turbines** have a radial inlet and axial outlet, such as the Francis turbine.

4. **Axial-flow turbines** have flow along the shaft axis, such as the Kaplan turbine.

Turbines are also classified into impulse turbines and reaction turbines:

- **Impulse turbines** change the direction of the flow of a high-velocity fluid jet. The resulting impulse spins the turbine and reduces the kinetic energy of the flow. There is no pressure change of the fluid in the turbine rotor blades. Before reaching the turbine, the fluid's pressure head is changed to a velocity head by accelerating the fluid with a nozzle. Pelton wheels use this process exclusively. Impulse turbines do not require a pressure casement around the runner, since the fluid jet is prepared by a nozzle prior to reaching the turbine.

- **Reaction turbines** develop torque by reacting to the fluid's pressure or weight. The pressure of the fluid changes as it passes through the turbine rotor blades. A pressure casement is needed to contain the working fluid as it acts on the turbine stage(s), or the turbine must be fully immersed in the fluid flow. The casing contains and directs the working fluid and, for water turbines, maintains the suction imparted by the draft tube. Francis and Kaplan turbines are different types of reaction turbine. For compressible working fluids such as gases, multiple turbine stages may be used to harness the expanding gas efficiently.

1.13   Summary

Although the connection between the tides and the Moon had been recognized by Aristotle, three centuries before Christ, Earth-centred astronomy persisted. Newton wrote the equilibrium theory in the late seventeenth century, and this was developed by Laplace who recognized that the response functions would have the same frequencies as the astronomical drivers. Kelvin took these forward, applied harmonic analyses, and introduced the naming convention for the constituents.

The harmonic current equation might be rewritten as

$$U(t) = U_0 + \sum \left[U_A \cos \left(\frac{2\pi t}{T_A} + \rho_A \right) \right] \tag{1.12}$$

The first two constituents are the lunar semi-diurnal constituent, with a symbol M_{U2} and a period of 12.4206012 h, and the solar semi-diurnal S_{U2}, with a period of 12.00 h.

1.14   Bibliography

AMS (2007) http://www.ams.org/featurecolumn/archive/tidesIII2.html [accessed 8 May 2007].

Andrews, J. & Jelley, N. (2007) *Energy Science: Principles, Technologies and Impacts*. OUP, Oxford, UK, 328 pp.

Art of Navigation (2007) http://www.abc.net.au/navigators/navigation/history.htm [accessed 14 November 2007].

Association d'Oceanographique Physique (1955) *Bibliography on Tides 1665–1939. Publication Scientifique 15*. Bergen, Norway. http://www.airmynyorks.co.uk/BibMod.html [accessed 1 May 2007].

Cartwright, D.E. (1999) *Tides: A Scientific History*. Cambridge University Press. Cambridge, UK, 292 pp.

Charlier, R.H. (2003) Sustainable co-generation from the tides: bibliography. *Renewable and Sustainable Energy Reviews* 7, 561–563.

Colepresse, S. (1668) Tides observed at Plymouth. *Phil. Trans. R. Soc. Lond.* 2, 632.

Defant, A. (1961) *Physical Oceanography*. Pergamon Press, Oxford, UK.

Drennon, W. (2007) http://home.cvc.org/science/kepler.htm [accessed 5 May 2007].

Electrical Engineering Tutorials (2008) http://powerelectrical.blogspot.com/2007/04/water-turbines-and-its-classification.html [accessed 7 March 2008].

ETMTL (2007) http://www.elingtidemill.wanadoo.co.uk/sitem.html [accessed 8 May 2007].

FWERI (2007) http://www.baw.de/vip/en/departments/department_k/methods/kenn/frqw/frqw-en.html#glo [accessed 7 May 2007].

Groves-Kirkby, C.J., Denman, A.R., Crockett, R.G.M., Phillips, P.S., & Gillmore, G.K. (2006) *Science of the Total Environment* 367, 191–202.

Hansen, R.D. (2007) *Water Wheels.* http://www.waterhistory.org/histories/waterwheels/[accessed 5 March 2008].

Kelvin (1911) *Mathematical and Physical Papers (Vol. VI)*, pp. 272–305 [from the *Minutes of the Proceedings of the Institution of Civil Engineers*, 11 March 1882.]

LEGOS (2008) http://www.legos.obs-mip.fr/en/[accessed 19 June 2008].

Moray, R. (1665) Extraordinary tides in the west Isles of Scotland. *Phil. Trans. R. Soc. Lond.* 1, 53.

Moray, R. (1666a) Considerations and inquiries concerning tides. *Phil. Trans. R. Soc. Lond.* 1, 298.

Moray, R. (1666b) Patterns of the tables proposed to be made for observing tides. *Phil. Trans. R. Soc. Lond.* 1, 311.

NOOA (2007a) http://www.co-ops.nos.noaa.gov/predhist.html [accessed 5 May 2007].

NOAA (2007b) http://tidesandcurrents.noaa.gov/predmach.html [accessed 5 May 2007].

Norwood, R. (1668) Of the tides at Bermuda. *Phil. Trans. R. Soc. Lond.* 2, 565.

Palmer, H. (1831) A description of a graphical register for tides and wind. *Proc. R. Soc. Lond.* 121, 209–213.

Pannikar, N.K. & Srinivasan, T.M. (1971) The concept of tides in ancient India. *Indian J. Hist. Sci.* 6, 36–50.

Philips, H. (1668) Time of the tides observed at London. *Phil. Trans. R. Soc. Lond.* 2, 656.

Phillips, A. (2007) *Harmonic Analysis and Prediction of Tides*. http://www.math.sunysb.edu/~tony/tides/harmonic.html [accessed 7 May 2007].

Proudman Oceanographic Laboratory (2007a) http://www.pol.ac.uk/home/insight/tideinfo.html [accessed 7 May 2007].

Proudman Oceanographic Laboratory (2007b) http://www.abc.net.au/navigators/navigation/history.htm [accessed 14 November 2007].

Pugh, D. (1996) *Tides, Surges and Mean Sea Level*. John Wiley & Sons, Ltd, Chichester, UK.

Ramage, J. (2004) Hydroelectricity, in *Renewable Energy*, 2nd edition, ed. by Boyle, G., Oxford University Press, Oxford, pp. 148–194.

Round, D.G. (2004) *Incompressible Flow Turbomachines*. Butterworth-Heinemann, Oxford, 352 pp.

Rouse, W.W. (1908) A *Short Account of the History of Mathematics*, 4th edition. Macmillan, London, UK.

Sarton, G. (1931) *Introduction to the History of Science, Vol. II, Part II*. Carnegie Institution of Washington, Washington, DC.

Smeaton, J. (1759) An experimental enquiry concerning the natural powers of water and wind to turn mills, and other machines, depending on a circular motion. *Phil. Trans. R. Soc. Lond.* 51, 100–174.

Stafford, R. (1668) Of the tides at Bermuda. *Phil. Trans. R. Soc. Lond.* 2, 792.

The Galileo Project (2007) http://galileo.rice.edu/sci/brahe.html [accessed 5 May 2007].

Tyson (2007) http://www.pbs.org/wgbh/nova/galileo/mistake.html [accessed 1 May 2007].

Wallis, J. (1666) Hypothesis on the flux and reflux of the sea. *Phil. Trans. R. Soc. Lond.* 1, 263–281.

Wallis, J. (1668) On the variety of the annual tides in several places of England. *Phil. Trans. R. Soc. Lond.* 2, 652.

2

Tidal hydraulics

2.1 Introduction

This chapter lays down the theoretical basis for tidal phenomena and for the fluid flow in tidal streams. Sections deal with the generalized theory of fluid flow, the concept of hydraulic power, and the Bernoulli and von Kármán equations, in preparation for the following chapters. The equations that govern the behaviour of the tide in deep and in shallow water are then presented, and attention focuses on the generalized solutions for tidal currents in open sea and at coastal and estuarine sites.

This chapter begins with some indication of the data that we are wishing to explain and ultimately to predict. Figure 2.1 shows a tidal current record obtained and displayed in real time from Cove Point in Chesapeake Bay. The data are available through the excellent North American NOAA Tides and Currents site

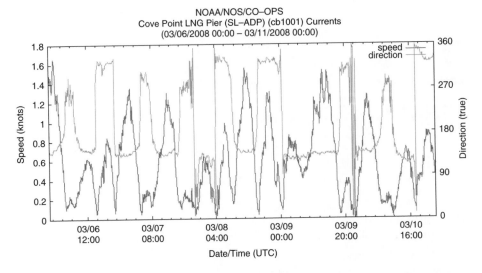

NOAA/NOS/CO–OPS
Cove Point LNG Pier (SL–ADP) (cb1001) Currents
(03/06/2008 00:00 – 03/11/2008 00:00)

Figure 2.1 Tidal current speed and direction at Cove Point in Chesapeake Bay on the east coast of North America (from NOAA, 2008).

which covers the work of the Centre for Operational Oceanographic Products and Services (COOPS).

The data cover the period from midnight on 5/6 March 2008 to midnight on 10/11 March 2008. The tidal currents are semi-diurnal, with peak flows occurring approximately every 6 hours (corresponding to both the flood and ebb currents), and achieve maximum speeds of around 1.5 knots ($0.75 \, \text{m s}^{-1}$).

The following Chapter is based upon a number of text books including Acheson (1990), ETSU (1993), Dyer (1997), Pugh (2004), Buiges et al (2006) and Andrews & Jelley (2007), as well as more specific papers referenced in the body of the text.

2.2 Elementary fluid flow

Consider the water channel shown in Figure 2.2. The channel becomes narrower and shallower in a downstream direction.

Three groups of terms are used to describe these flows. Firstly, the flow in the wider upper section is clearly slower than the flow in the narrower

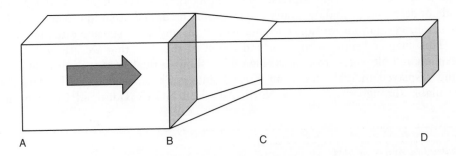

Figure 2.2 Schematic representation of the flow though a converging channel.

lower section, and the flow accelerates through the changing cross-sections. By definition:

- **Uniform flow** has a constant velocity with distance (A–B and C–D)

- **Non-uniform flow** has a variable velocity with distance (B–C).

Secondly:

- **Steady flow** is constant through time (for example, a river system).

- **Unsteady flow** varies through time (for example, a tidal system).

Thirdly, a surface wave travels at a speed that depends upon the square root of the water depth. The ability of a surface wave to travel upstream, against the current, thus depends upon the water depth. This provides another set of descriptive terms:

- **Subcritical flow** exists when the surface wave is able to progress upstream against the current.

- **Critical flow** exists when the surface wave travels upstream with the same celerity as the current.

- **Supercritical flow** exists when the surface wave is unable to travel upstream against the current and is swept downstream.

In general, tidal flows accelerate, decelerate, and reverse over varying depths and are therefore unsteady and non-uniform and can vary from subcritical, through critical, to supercritical within each tidal cycle. Since the volume of fluid passing through A–B, Q ($m^3\ s^{-1}$), must be identical to the volume of fluid passing through C–D, it follows that

$$Q = w_1 d_1 U_1 = w_2 d_2 U_2 \tag{2.1}$$

where U_1 and U_2 are the velocities, w_1 and w_2 are the widths, and d_1 and d_2 are the depths at section (A–B) and (C–D) respectively. Hence

$$U_2 = \frac{w_1 d_1 U_1}{w_2 d_2} \tag{2.2}$$

2.3 Hydraulic power

The hydraulic power density in flowing water is equal to the kinetic energy of the fluid that passes through a unit area of the flow normal to the dominant flow direction in unit time:

$$K_E(t) = \frac{1}{2} M U(t)^2 \tag{2.3}$$

The relevant mass is equal to the product of the density of water and a volume of unit length in the vertical and crossflow directions and of length equivalent to the downstream velocity in the third dimension:

$$M = \rho U(t) \tag{2.4}$$

Thus, the hydraulic power density is given by combining equations (2.3) and (2.4):

$$P_D(t) = \frac{1}{2}\rho U(t)^3 \tag{2.5}$$

2.4 Turbulence and the Reynolds number

Materials such as water flow in response to applied stress. Different materials respond in different ways to the same amount of applied stress. If, for example, an eraser, a ball of clay, a cube of salt, and a cubic centimetre of honey are dropped onto the floor, each responds differently. The eraser rebounds, the soft clay sticks, the salt crystal fractures, and the honey spreads out slowly. These differences in behaviour illustrate elastic, plastic, fracture, and viscous deformation respectively (Table 2.1). Elastic deformation is completely recoverable, whereas viscous and plastic substances cannot reverse the deformation:

Viscous substances such as water respond to an applied stress by flowing, and the rate of deformation (flow) depends on the magnitude of the stress. This is summarized by the fundamental relationship

$$\tau = \mu \frac{du}{dz} \tag{2.6}$$

The shear stress is proportional to the velocity gradient du/dz, where the coefficient of proportionality is the dynamic viscosity. Materials that have such a direct relationship between the stress and the velocity gradient are known as Newtonian fluids. Water is a typical Newtonian fluid and has a dynamic viscosity of about $0.001\,\text{kg m}^{-1}\,\text{s}^{-1}$. The resistance to applied stress that occurs at the molecular level is known as the *kinematic viscosity*, v, and is equal to the dynamic viscosity divided by the mass density.

The behaviour of water in an open tidal channel is governed by the effects of three forces on the fluid. These are the inertial, the viscous and the gravitational

Table 2.1 Response of different substances to applied stress

Material	Response	Deformation	Recovery
Eraser	Bounces	Elastic	Complete
Clay	Sticks	Plastic	Incomplete
Halite	Breaks	Fracture	None
Honey	Spreads out	Viscous	Flow

forces. Much of the fluid's behaviour depends upon the destabilizing effect of the inertial force compared with the stabilizing viscous force. In general, the inertial force is a function of the density of the fluid. In accordance with Newton's first law (Chapter 1), the greater the density of a fluid, the greater is the force required to produce a specified acceleration in a specified volume.

This phenomenon was studied by Osborne Reynolds (Figure 2.3), who carried out the experiment depicted in Figure 2.4 in the nineteenth century in Manchester, England. Reynolds was a British fluid dynamics engineer who was born in Belfast, Ireland, and died in Somerset, England. He graduated from Cambridge University in 1867 after studying mathematics and became a Professor of Engineering at Owens College, Manchester (a predecessor to the University of Manchester), where he worked until he retired in 1905.

In the Reynolds experiment a reservoir supplied water into a long glass tube within which a streak of dye entered at one end. A valve controlled the rate of water flow. Reynolds noted that, if the velocity was increased, the streak of dye in the water changed from a coherent and straight line into a complex series of vortices and lateral mixings. The coherent flow represents *laminar* conditions, where the stabilizing viscous forces dominate the inertial forces. The chaotic flow at higher speeds represents *turbulent* conditions, where inertial instabilities dominate. In laminar flow, a transfer of momentum from the faster to the slower water occurs at the molecular level, whereas in turbulent flow whole packets of water are transferred. Equation (2.6) applies directly to laminar flows, but for turbulent flows it is modified to

$$\tau = K_z \frac{\mathrm{d}u}{\mathrm{d}z} \tag{2.7}$$

The eddy viscosity, K_z, is several orders larger than the molecular viscosity but has the same dimensions.

Figure 2.3 Osborne Reynolds (1842–1912).

Figure 2.4 The Reynolds experiment.

It is often convenient to express the boundary shear stress in a form that has the dimensions of velocity. This is done by defining a friction velocity, U_*, such that

$$U_* = \sqrt{\frac{\tau}{\rho}} \tag{2.8}$$

The **inertial** force acting on a particle of fluid is given by Newton's law (Chapter 1) as equal in magnitude to the mass of the particle multiplied by its acceleration (Hardisty, 2007). The mass is equal to the density, ρ, multiplied by the volume (which is the cube of a length, L, characteristic of the geometry of the system). The mass is thus proportional to ρL^3. The acceleration of the particle is the rate at which its velocity changes with time and is thus proportional to the velocity divided by the time period. The time period may, however, be taken as proportional to the chosen characteristic length divided by the characteristic velocity, U (m s^{-1}), so that the acceleration may be set proportional to $U/(L/U) = U^2/L$. The magnitude of the inertial force is thus

$$\text{Inertial force} = \frac{\rho L^3 U^2}{L} = \rho L^2 U^2 \tag{2.9}$$

The **viscous force** is given by equation 2.6 which is proportional to $\mu U/L$. The magnitude of the area over which this force acts is proportional to L^2, and thus the magnitude of the viscous force per unit area is

$$\text{Viscous force} = \frac{\mu U}{L} L^2 = \rho UL \tag{2.10}$$

Finally, the **gravity force** on the particle of water is equivalent to the particle weight, which is its mass multiplied by the gravitational acceleration. The magnitude of the gravity force is thus

$$\text{Gravity force} = \rho L^3 g \tag{2.11}$$

The three ratios of a combination of two of these three forces representing the stabilizing to the destabilizing force are expressed as three non-dimensional numbers that characterize tidal waters. The three non-dimensional numbers are called the Reynolds, Froude, and Richardson numbers respectively.

The ratio of inertial forces to viscous forces determines whether the flow is laminar or turbulent, and is represented by the **Reynolds number**, Re. The Reynolds number is thus the ratio of equation (2.9) and equation (2.10) with the substitution $\nu = \mu/\rho$:

$$Re = \frac{UL}{\nu} \tag{2.12}$$

The Reynolds number is dimensionless and can be a measure of any size of system. It is independent of the units of measurement. For small values of Re, viscous forces prevail and flow is laminar, but, as Re increases, turbulence sets in. The values of Re that separate laminar and turbulent flow are called the transition zone and begin in the vicinity of Reynolds numbers of about 500–600. The upper boundary can be regarded as about 2000. Laminar flow rarely occurs in tidal channels except near the boundary.

2.5   Critical flow and the Froude number

The ratio of the destabilizing inertial forces to the stabilizing gravity force is also a measure of flow characteristics. The ratio is known as the **Froude number**, Fr, after William Froude (Figure 2.5), a pioneer in the study of naval architecture, who first introduced it. Nowadays the Froude number is usually taken as the square root of the ratio of the two forces given by combining equation (2.9) and equation (2.11):

$$Fr = \sqrt{\frac{\rho L^2 u^2}{\rho L^3 g}} = \frac{u}{\sqrt{gh}} \tag{2.13}$$

The characteristic length scale is replaced by the water depth h. The Froude number thus also represents the ratio of flow velocity to the speed of movement of gravity waves. When $Fr > 1$, the flow is said to be *supercritical* and a disturbance can travel upstream; when $F_r < 1$, it is *subcritical* and the disturbance is swept downstream (cf. Section 2.2). Critical flow conditions are represented by $Fr = 1$ and are discussed in Chapter 3.

Figure 2.5　William Froude (1810–1879).

2.6　Bernoulli's equation

Consider the steady flow of water in the control volume shown in Figure 2.6. The height, cross-sectional area, speed, and pressure at any point are denoted by z, A, u, and p respectively. The increase in gravitational potential energy of a mass δm of fluid between z_1 and z_2 is $\delta mg(z_1 - z_2)$. In a small time interval δt, the mass of fluid entering the control volume at A_1 is $\delta m = \rho u_1 A_1\, \delta t$ and the mass exiting at A_2 is $\delta m = \rho u_2 A_2\, \delta t$.

In order for the fluid to enter the control volume, it has to do work to overcome the pressure p_1 exerted by the fluid. The work done by pushing an elemental mass δm a short distance $\delta s_1 = u_1 \delta t$ at P$_1$ is

$$\delta W_1 = p_1 A_1\, \delta s_1 = p_1 A_1 u_1\, \delta t \tag{2.14}$$

Similarly, the work done in pushing the elemental mass out of the control volume at P$_2$

$$\delta W_2 = -p_2 A_2\, \delta s_2 = p_2 A_2 u_2\, \delta t \tag{2.15}$$

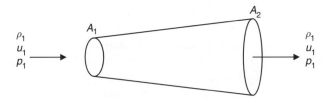

Figure 2.6　Control volume for the Bernoulli equation.

where the change of sign recognizes the negative work done. The net work done is $\delta W_1 + \delta W_2$:

$$\delta W_1 + \delta W_2 = p_1 A_1 \, \delta s_1 = p_1 A_1 u_1 \, \delta t - p_2 A_2 \, \delta s_2 = p_2 A_2 u_2 \, \delta t \tag{2.16}$$

By the principle of conservation of energy this is equal to the increase in kinetic energy plus the increase in potential energy, so that:

$$p_1 A_1 u_1 \, \delta t - p_2 A_2 \, \delta s_2 = p_2 A_2 u_2 \, \delta t = \delta mg(z_1 - z_2) + \delta m(u_2^2 - u_1^2) \tag{2.17}$$

Substituting

$$\delta m = \rho u_1 A_1 \, \delta t = \rho u_2 A_2 \, \delta t \tag{2.18}$$

yields

$$\frac{p_1}{\rho} + gz_1 + \frac{1}{2}u_1^2 = \frac{p_2}{\rho} + gz_2 + \frac{1}{2}u_2^2 \tag{2.19}$$

Finally, since points P_1 and P_2 are arbitrary, it follows that

$$\frac{p}{\rho} + gz + \frac{1}{2}u^2 = \text{const} \tag{2.20}$$

Equation (2.20) is known as the Bernoulli equation after the Swiss mathematician Jacob (Jacques) Bernoulli (1654–1705) (Figure 2.7), who was also the first to use the term 'integral', was an early user of polar coordinates, and discovered the isochrone. The significance of the Bernoulli equation is that it shows how the pressure in a moving fluid decreases as the flow accelerates. In the present context, it controls

Figure 2.7 Jacob Bernoulli (1654–1705).

flow through the ducts that are used in some tidal stream power devices to increase the efficiency of the rotor or turbine (Chapter 3).

2.7   von Kármán's equation

The fluid force exerted on the bed of the tidal channel within the boundary layer depends upon both the flow velocity and the roughness or drag of the bed (Hardisty, 2007) and is derived from the shear of the flow (equations (2.6) and (2.7)). The boundary layer is the layer of reduced velocity in fluids, such as water, that is immediately adjacent to the surface of the solid past which the fluid is flowing. The simplest determination of the shearing stress that the flow exerts on the sediment is known as the quadratic stress law (Lewis, 1997) (Chapter 3):

$$\tau = \rho C_{D100} u_{100}^2 \tag{2.21}$$

The coefficient of proportionality is known as the Drag coefficient (Miller et al., 1982, 1983, 1984 and 1985)

This expression is valid for a flow in which the Reynolds number is sufficiently high for friction to depend upon the roughness of the surface so that the influence of viscosity is negligible. Dyer (1986) quotes the values for C_{D100} shown in Table 2.2.

An alternative analysis of the shear stress on the bed was developed by Theodore von Kármán (Figure 2.8), who was a Hungarian–American engineer and physicist active primarily in the fields of aeronautics and astronautics. von Kármán introduced a non-dimensional constant describing the logarithmic velocity profile of a turbulent fluid near a boundary. The equation for such boundary layer flow profiles is

$$U = \frac{U_*}{\kappa} \ln \frac{z}{z_0} \tag{2.22}$$

The von Kármán constant, κ, typically takes the value 0.41. Equation (2.22) represents von Kármán's law of the wall, which states:

> There is a linear relationship between the flow velocity in a turbulent boundary layer and the logarithm of the height above the bed.

Table 2.2 Drag coefficients for typical seabed types, based on a velocity measured 100 cm above the bed (after Dyer (1986) and Soulsby (1983))

Bottom type	C_{D100}	Bottom type	C_{D100}
Silt/sand	0.0014	Mud	0.0022
Sand/shell	0.0024	Sand/gravel	0.0024
Mud/sand/gravel	0.0024	Unrippled sand	0.0026
Mud/sand	0.0030	Gravel	0.0047
Rippled sand	0.0061		

Figure 2.8 Theodore von Kármán (1881–1963).

The velocity at any height can be determined from equation (2.22). The intercept, z_0, is called the roughness height at which the velocity becomes zero and has the typical values shown in Table 2.3.

Measurements have verified that the von Kármán profile applies in the sea (Dyer, 1986) and may thus be employed to determine the vertical flow variations across a tidal power device. This was demonstrated, for example, by Charnock (1959), Sternberg (1966), Dyer (1970) and Channon and Hamilton (1971). Some aspects of the problem have been considered by Dyer (1986)who demonstrated that, provided measurements were averaged for a period in excess of 20 times the characteristic timescale of the flow, no further improvement in the semi-logarithmic profile resulted from longer time averages. In practice this requires averaging over about 10 min.

Away from the bed, the relative importance of the inertia to viscous effects is greater than near the bed. Consequently, in an accelerating current the flow well away from the boundary will retain a 'memory' of the preceding driving forces for a longer time than near the bed. The flow profile will therefore be concave upwards and the current speed smaller than the semi-logarithmic value by an amount that increases with z. Conversely, in a decelerating flow, the profile will be concave downwards.

Table 2.3 Roughness heights for typical seabed types (after Dyer (1986) and Soulsby (1983))

Bottom type	$z_0 m \times 10^{-3}$	Bottom type	$z_0 m \times 10^{-3}$
Silt/sand	0.02	Mud	0.2
Sand/shell	0.3	Sand/gravel	0.3
Mud/sand/gravel	0.3	Unrippled sand	0.4
Mud/sand	0.7	Gravel	3.0
Rippled sand	6.0		

An alternative description of the current velocity profile in the boundary layer was developed by the Dutch water engineer Johan van Veen, architect of the delta plan of sea defences across the Rhine. In the course of preliminary surveys in the late 1930s, van Veen measured the profiles of tidal currents from anchored ships at a large number of stations in the southern North Sea and the Dover Strait. The profiles varied with location, but van Veen (1938) found that an empirical law

$$U = U_h \frac{z}{h}^{1/\kappa_V} \tag{2.23}$$

(with κ_V in the region of 5.2) gave a remarkably close fit to most of his profiles.

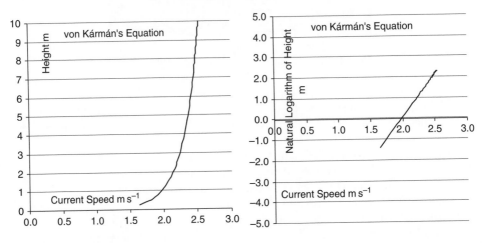

Figure 2.9 von Kármán's equation for $U_* = 1.0\,\mathrm{m\,s^{-1}}$ and a mud/sand/gravel mix with $z_0 = 0.0003\,\mathrm{m}$ (from online models, cf. Introduction) with linear height (left) and logarithmic height (right).

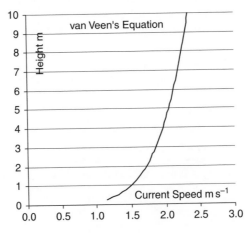

Figure 2.10 van Veen's equation for $U_h = 1.5$ at $h = 1.0$ and $\kappa_V = 5.2$ (from online models, cf. Introduction).

Examples of the von Kármán and van Veen profiles are shown in Figures 2.9 and 2.10, demonstrating that there is little difference between the two approaches. We shall return to the effect of these velocity profiles in defining the hydraulic power and thus the efficiency of tidal stream power devices in later chapters.

2.8 Properties of the Airy wave

The tide behaves as a gravity wave in that the gravitational force tends to oppose the sea surface disturbances. There are a number of mathematical theories for gravity waves such as Stokes, cnoidal, and solitary waves (Hardisty, 1992), but the simplest is the linear or Airy wave theory, which is utilized here.

George Airy (Figure 2.11) entered Trinity College, Cambridge, in 1819, where his performance was outstanding. He graduated as the top First Class student in 1823 and was awarded a fellowship at Trinity College. Only 3 years later, he was appointed Lucasian Professor of Mathematics at Cambridge. Airy developed the small-amplitude wave theory which applies, reasonably well, to tides in the open sea, but not to the estuarine environments discussed in the following section. The surface profile of the wave can be considered to be a progressive sinusoid with amplitude equal to half the tidal range. The wave motion is described by the wave dispersion equation (Dyer, 1986; Hardisty, 1992):

$$\sigma^2 = gk \tanh kh \tag{2.24}$$

An individual wave then has phase celerity:

$$c = \frac{\lambda}{P} = \left(\frac{g\lambda}{2\pi} \tanh \frac{2\pi h}{\lambda} \right)^{1/2} = \frac{gT}{2\pi} \tanh \frac{2\pi h}{\lambda} \tag{2.25}$$

Figure 2.11 George Airy (1801–1892).

In practice it is comparatively easy to measure the period of the tide and the water depth, but the wavelength is much more difficult. If the period and depth are known, then the wavelength can be calculated iteratively from equation (2.24), as described, for example, by Hardisty (1992).

In shallow water, the ratio h/λ is small and the hyperbolic tangent, $\tanh 2\pi h/\lambda$, approaches $2\pi h/\lambda$ so that

$$c = \sqrt{gh} \qquad (2.26)$$

Equation (2.26) is a fundamental relationship that we have already encountered in discussion of critical flow and the Froude number.

2.9   Tides in estuaries

Tides in the open ocean and offshore generally behave in accordance with the Airy wave theory described above. The surface profile is essentially sinusoidal, or can be described by equation (1.11), and the corresponding currents can be described by equation (1.12). The two end-members in the continuum of surface wave characteristics are called progressive waves and standing waves. The tide in the deep ocean, far from land, has the characteristics of a progressive wave, whereas, by the time it arrives at the upper reaches of an estuary and is reflected from the banks and bars, it has the characteristics of a standing wave.

Flow velocities in a progressive wave are 'in phase' with water depth, as shown in Figure 2.12 (upper), with a flood that reaches a maximum speed close to high water and an ebb that reaches a maximum speed close to low water.

Within the confines of an estuary, however, the wave is reflected and exhibits a behaviour that is generally known as a standing wave (Hardisty, 2007), with the flow and water depths out of phase, as shown in Figure 2.12 (lower). In an estuary

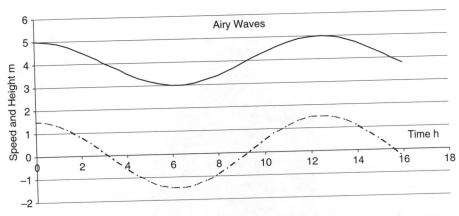

Figure 2.12 The water depth (upper solid line) and corresponding Airy wave flow velocities (lower, broken line and landwards positive) for (upper) progressive tidal waves and (lower) standing tidal waves (from online models, cf. Introduction).

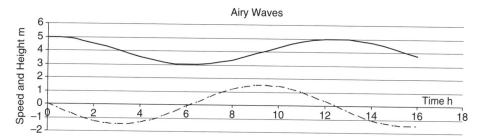

Figure 2.12 (*continued*)

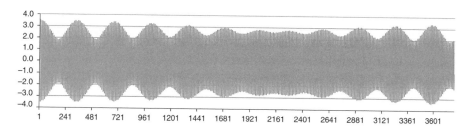

Figure 2.13 Absolute value of tidal current velocity (m s⁻¹) computed at hourly intervals for the Hull Roads site described in Section 2.13 for a 6 month interval.

the tide turns at low water and the flood runs landwards, reaching a maximum at mid-tide and then decreasing until the tide turns seawards at high water. The ebb reaches a maximum at mid-tide, but this time on a falling tide until the next low water is reached.

2.10   Harmonic prediction of tidal currents

Tidal changes in water level are fed by very large horizontal movements of water having the same cyclic periodicity (cf. Chapter 1). Where the horizontal water motions are particularly pronounced, they are referred to as *tidal currents*. The intensities of the horizontal flows are very sensitive to the sizes of sea areas, the variations in water depths (bathymetry), and the shapes of land masses.

Figure 2.13 shows the change in tidal current velocities throughout a year for a site in the Fall of Warness (Orkney). The peaks during spring tides are clearly visible.

In restricted areas, such as channels and narrow estuaries, tidal currents are more or less bidirectional. In more open areas of the sea, the direction of flow (which may be plotted in the form of a *tidal ellipse*) may be complicated by factors that include the Coriolis force which diverts large ocean currents to the right in the Northern Hemisphere and to the left in the Southern Hemisphere. Tidal currents are sensitive to bathymetric changes and to bottom friction. In general, current velocities reduce with depth, particularly near to the seabed (cf. Section 2.7). Weather conditions may also have some additional influence. Wind can reinforce or weaken tidal motions, and water level is influenced by changes in atmospheric pressure.

Table 2.4 Principal species used in the simulation of tidal currents in the STEM model detailed in later chapters with respect to the MSD and then hours, with the MSD set to $365.2422 \times 24\,$h

Symbol	Name	Period (h)
M_{U2}	Principal lunar semi-diurnal constituent	12.4206
S_{U2}	Principal solar semi-diurnal constituent	12.0000
M_{U4}	Lunar quarter diurnal	6.2103
O_{U1}	Lunar diurnal constituent	25.8200
K_{U1}	Lunisolar diurnal constituent	23.9300
K_{U2}	Lunisolar semi-diurnal constituent	11.9650

The tidal currents may be predicted from knowledge of the component harmonics. Equation (1.12) from Chapter 1 can be expanded (equation (2.27)) for the first six tidal constituents that are utilized in later chapters for the simplified tidal economic model (STEM):

$$
\begin{aligned}
U(t) = U_0 &+ U_{MU2} \cos\left[\frac{2\pi t}{T_{MU2}} + \rho_{MU2}\right] + U_{SU2} \cos\left[\frac{2\pi t}{T_{SU2}} + \rho_{SU2}\right] \\
&+ U_{MU4} \cos\left[\frac{2\pi t}{T_{MU4}} + \rho_{MU4}\right] + U_{OU1} \cos\left[\frac{2\pi t}{T_{OU1}} + \rho_{OU1}\right] \\
&+ U_{KU1} \cos\left[\frac{2\pi t}{T_{KU1}} + \rho_{KU1}\right] + U_{KU2} \cos\left[\frac{2\pi t}{T_{KU2}} + \rho_{KU2}\right]
\end{aligned}
\tag{2.27}
$$

These constituents are shown in Table 2.4 and are detailed in later chapters. Note that the tidal current harmonic corresponding to the tidal amplitude harmonic is signified by the addition of subscript U in the designation. Thus, M_{U2} corresponds to M_2.

2.11   Turbulent characteristics of tidal currents

Measurements of the velocity vectors in a turbulent tidal current that are averaged over a sufficiently long period of time will give a very much more consistent result than the instantaneous data. The random fluctuations become averaged out with time, and a steady background current can be distinguished that moves the turbulence along and on which the turbulence is superimposed. The instantaneous velocities can be separated into three orthogonal components, U, V, and W, which act respectively in the x streamwise horizontal direction, the y lateral horizontal direction, and the z vertical upwards direction. The horizontal velocity U at a point and at any instant in time can be considered as being composed of a time mean flow, U_M, and a turbulent deviation, U'. Thus

$$U = U_M + U' \tag{2.28}$$

$$V = V_M + V' \tag{2.29}$$

$$W = W_M + W' \tag{2.30}$$

The mean of the turbulent deviation is, of course, zero, but the mean of the deviation squared is not zero. The root mean squared values are usually normalized against the mean flow in order to be able to compare magnitudes at different levels or in different flows. The parameter is called the turbulence intensity, I_T (Dyer, 1986):

$$I_T = \frac{\sqrt{\overline{U'^2}}}{U_M} \qquad (2.31)$$

Near the boundary, the turbulence intensity is generally of the order of 10 %, but it decreases away from the wall. Although the mean of the instantaneous turbulent velocity is equal to the mean current, the mean of the hydraulic power density (Section 2.2), which is related to the cube of the current speed, is not equal to the hydraulic power density at the mean current speed, but is proportionally higher. A more accurate measurement is to take the average of the maximum and the minimum hydraulic power density. If, for example, the average maximum instantaneous current is $1.1U_M$ and the minimum instantaneous current is $0.9U_M$, so that the mean hydraulic power density (equation (2.6)) as a ratio to the hydraulic power density at U_M, here called the intensity of the turbulent power, I_{TP}, is

$$I_{TP} = 500\frac{\left[(1.1U_M)^3 + (0.9U_M)^3\right]}{1000U_M^3} = 1.03 \qquad (2.32)$$

Thus, the hydraulic power density is some 3 % higher if 10 % turbulent fluctuations are included than would be indicated by the mean water speed. Although this may appear small, it is very significant in the context of the commercialization of tidal power devices, where a few percentage points in operating efficiency may cover the difference between a profitable venture and a financial loss. These issues are discussed further in later chapters.

The turbulent eddies responsible for these instantaneous deviations have a whole variety of length and timescales. Near the seabed or the surface, the eddies must have restricted vertical dimensions, but their horizontal scales can still be fairly large. The largest eddies, however, can have a vertical scale equivalent to the water depth and a horizontal scale equivalent to the dimensions of the channel. The small eddies are generated by bed shear, and they rise into the body of the flow as they are translated downstream. Large eddies are generated by lateral shear against gross boundary changes such as headlands or channel meanders. They lose energy by interacting to create smaller eddies and become larger and slower with distance. Eventually the interacting eddies will be broken down to scales where the energy can be dissipated as heat by viscosity. The cascade of energy from larger scales to smaller scales has been aptly described in Warhaft (1997) as:

> Big whirls have little whirls,
> Which feed on their velocity,
> Little whirls have lesser whirls,
> And so on unto viscosity.

Water movements in the marine environment form a spectrum. The large, lower-frequency eddies are the result of motions of tidal period that have very large energies. This energy is passed down to smaller eddies by the cascade process without a significant energy loss. The gradient of the spectrum in this region is proportional to the frequency raised to the power 5/3. There is also the input of turbulent tidal energy at periods of around 1–100 s from the turbulent bed shear.

2.12 Redundancy and intermittency

Most forms of renewable energy are unreliable – the wind does not blow all of the time, for example, and this is taken into account when analysing the financial viability of the project. With the exception of hydroelectricity, which is fairly uniform in its power-generating capability, electricity generation from other renewable energy sources is usually argued to be both intermittent and unpredictable (Panmure Gordon, 2006). This means that a high level of redundancy will be required. For example, the UK grid system could not depend on wind generation to satisfy peak load because, quite simply, the wind might not be blowing at that time. As renewable energy forms a larger part of our requirements, the costs of backing up this generation could therefore become significant. Tidal power offers the opportunity to address this problem owing to the varying phase relationships around coastlines. The Atlantic semi-diurnal tidal wave travels from south to north. Energy is transmitted across the shelf edge into the Celtic Sea between France and southern Ireland. This wave then propagates into the English Channel where some energy leaks into the southern North Sea. It also spreads into the Irish Sea and the Bristol Channel. The Atlantic wave travels northwards, taking about five hours to travel from the Celtic Sea to the north of Scotland. The semi-diurnal wave is partly diffracted off the north coast of Scotland where it turns to the east and to the south into the North Sea. The semi-diurnal tides of the North Sea consist of two complete amphidromic systems (cf. Chapter 7), centred in the German Bight and off Suffolk, and a third, degenerate system centred in southern Norway.

We now analyse the installation of tidal power stations around the coast of the United Kingdom in order to provide continuous power and thus obviate the intermittency problem. After consideration of the phase differences for the semi-diurnal tidal currents around the UK coastline, six sites were chosen at which the current speeds are about two hours out of phase as the tidal wave travels northwards up the west coast and then southwards down the east coast. The sites are shown in Table 2.5. UKHO (2002, 2003, and 2004) was used to determine the relative time of the tides at each of these sites (for the morning tides of Tuesday 21 November 2006), and these are listed in the table. The times were then corrected to represent a phase delay or time lag in the tides with respect to a zero hour taken to be in the Severn as shown in the table. UKHO (2004) was used to determine a maximum spring tidal current for each

Table 2.5 The sum of the total installed capacities is 10.7 MW, and the sum of the total annual outputs is 18 800 GW h. The mean of the hourly totals of the outputs is 4.6 MW

	Severn	Menai St	Mersey	Clyde	Tyne	Humber
Tidal diamond	SN052L	SN048F	SN045M	SN040A	SN0201	SN017A
Latitude	51 23.23	53 08.72	53 23.02	55 58.59	55 00.91	53 43.85
Longitude	03 04.98	04 16.67	02 59.78	04 44.07	01 22.20	00 20.92
Time	7.10	9.80	11.10	1.00	3.20	5.75
Phase (h)	0.00	2.70	4.00	5.90	8.10	10.65
Peak current (knots)	3.9	3.6	5.2	4.2	2.6	5.0
Peak current (m s^{-1})	1.95	1.80	2.6	2.1	1.3	2.5
Maximum power (kW unit^{-1})	148	117	352	185	44	313
Mean power (kW unit^{-1})	30	23	70	37	9	63
Annual output (MW h unit^{-1})	260	204	616	325	77	548
Balancing units	**17**	**10**	**5**	**5**	**24**	**5**
Installed capacity (MW)	2.5	1.2	1.8	0.9	2.8	1.6
Annual output (MW h)	4417	2044	3079	1623	4904	2738

site. It should be noted that this flow information is only indicative as it represents Admiralty tidal diamond information and may not be from the strongest or most suitable location. It should also be noted that the peak spring current is not the maximum current at the site; it is the maximum current on a mean spring tide.

The peak spring currents listed were then used to calculate a maximum power output or capacity for the device from the product of the potential hydraulic power density, the capture area A (m^2), which was arbitrarily taken to be 100 m^2, and an efficiency e, which is taken as 40 %. The mean annual output power from these hypothetical units at each site is calculated by multiplying the installed capacity by the capacity factor, which it was argued is to be taken as 20 %. The results are shown in Table 1, indicating that the annual output from each hypothetical power unit ranges from less than 100 MW h in the Tyne to more than 600 MW h in the Mersey.

A short computer routine was written to simulate the phase-locked currents from the six sites and to calculate the total power generated. The number of units 'installed' at each site (the balancing units in Table 2.5) was varied until, for the values shown in the table, there was a steady national generation without interruption of about 4.6 MW (Figure 2.14). The installed capacity and annual power generation are also shown in the table. The total for all units at all sites is an installed capacity of about 10.7 MW and an annual production of about 18.8 GW h. The data given, however, are only examples, and, with the proper consents and financial support, many more units could be installed, although this may not be practical in some estuaries.

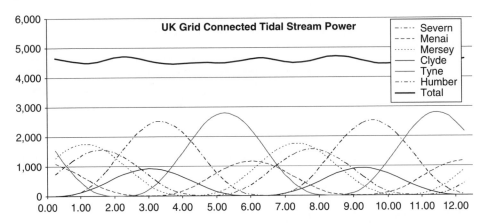

Figure 2.14 Power generation (kW) at each of the six phase-locked sites and the total if all sites are grid connected throughout a typical 12 h period.

2.13 Summary

This chapter has demonstrated that flow may be uniform or non-uniform, steady or unsteady, and subcritical, critical, and supercritical, and that the velocity flowing out of a cross-section is given by

$$U_2 = \frac{w_1 d_1 U_1}{w_2 d_2} \qquad (2.2)$$

The hydraulic power density is given as

$$P_D(t) = \frac{1}{2}\rho U(t)^3 \qquad (2.5)$$

Flow can also be laminar or turbulent, where the onset of turbulence corresponds to a Reynolds number (the ratio of the inertial to the viscous forces) of about 500–2000:

$$Re = \frac{UL}{\nu} \qquad (2.12)$$

The kinematic viscosity has a value of about 10^{-6}. The ratio of the inertial to the gravity force is the Froude number:

$$Fr = \frac{u}{\sqrt{gh}} \qquad (2.13)$$

When $Fr > 1$, the flow is said to be *supercritical* and a disturbance can travel upstream; when $Fr < 1$, it is *subcritical* and the disturbance is swept downstream. The Bernoulli equation relates the kinetic and pressure energies within a flow:

$$\frac{p}{\rho} + gz + \frac{1}{2}u^2 = \text{const} \tag{2.20}$$

von Kármán's equation describes the boundary layer within a flow:

$$U = \frac{U_*}{\kappa} \ln \frac{z}{z_0} \tag{2.22}$$

where the shear velocity is given by

$$U_* = \sqrt{\frac{\tau}{\rho}} \tag{2.8}$$

The tide can be described approximately by linear (Airy) wave theory:

$$h(t) = H_0 + H \cos\left[\frac{2\pi t}{T} + \rho_h\right] \tag{1.11}$$

$$U(t) = U_0 + U_A \cos\left[\frac{2\pi t}{T_A} + \rho_A\right] \tag{1.12}$$

If the phase difference of the surface profile is set to zero, tidal currents offshore are progressive and have a phase difference also of zero. In estuaries the tidal currents are symptomatic of a standing wave and have a phase difference of $T/4$. In more detail, the tidal currents can be simulated in terms of the first six harmonics by

$$\begin{aligned} U(t) = U_0 &+ U_{MU2} \cos\left[\frac{2\pi t}{T_{MU2}} + \rho_{MU2}\right] + U_{SU2} \cos\left[\frac{2\pi t}{T_{SU2}} + \rho_{SU2}\right] \\ &+ U_{MU4} \cos\left[\frac{2\pi t}{T_{MU4}} + \rho_{MU4}\right] + U_{OU1} \cos\left[\frac{2\pi t}{T_{OU1}} + \rho_{OU1}\right] \\ &+ U_{KU1} \cos\left[\frac{2\pi t}{T_{KU1}} + \rho_{KU1}\right] + U_{KU2} \cos\left[\frac{2\pi t}{T_{KU2}} + \rho_{KU2}\right] \end{aligned} \tag{2.27}$$

The turbulent intensity of a tidal current is given by

$$I_T = \frac{\sqrt{\overline{U'^2}}}{U_M} \tag{2.31}$$

which may be combined with equation (2.5) to show that 3 % more energy is generated from the instantaneous currents at $I_T = 0.1$ than from the mean, steady currents at $I_T = 0$.

Finally, it is possible to connect tidal stream power outputs from estuarine locations around the UK to produce a steady electrical power due to the tidal lags between sites.

2.14 Bibliography

Acheson, D.J. (1990) *Elementary Fluid Dynamics*. Oxford University Press, Oxford, UK, 397 pp.

Andrews, J. & Jelley, N. (2007) *Energy Science: Principles, Technologies and Impacts*. Oxford University Press, Oxford, UK, 328 pp.

[accessed 7 May 2007].

Buigues, G., Zamora, I., Mazon, A.J., *et al.* (2006) *Sea Energy Conversion: Problems and Possibilities*. International Conference On Renewable Energies And Power Quality, 2006.

Channon, R.D. & Hamilton, D. (1971) Sea bottom velocity profiles on the continental shelf south-west of England. *Nature* **21**, 383–385.

Charnock, H. (1959) Tidal friction from currents near the seabed. *Geophys. J. R. Astronomical Soc.* **2**, 215–221.

Dyer, K.R. (1970) Current velocity profiles in a tidal channel. *Geophys. J. R. Astronomical Soc.* **22**, 153–161.

Dyer, K.R. (1986) *Coastal and Estuarine Sediment Dynamics*. John Wiley & Sons, Ltd, Chichester, UK, 342 pp.

Dyer, K.R. (1997) *Estuaries: A Physical Introduction*. John Wiley & Sons, Ltd, Chichester, UK, 195 pp.

Hardisty, J. (1992) *Beaches: Form and Process*. Unwins, London, UK.

Hardisty, J. (2007) *Estuaries: Monitoring and Modelling the Physical System*. Blackwells, Oxford, UK.

Hydrographic Office (2004) *Total Tide*. Tidal Prediction Software. Available as DP550-2004 from http://www.outdoorgb.com/p/admiralty_windows_tidal_prediction_software/

Lewis, R., (1997). *Dispersion in Estuaries and Coastal Waters*. John Wiley & Sons, Ltd, Chichester, UK, 312pp.

Miller, G., Corren, D., & Armstrong, P. (1984) Kinetic hydro energy conversion systems study for the New York State resource – Phase II and Phase III model testing. New York Power Authority.

Miller, G., Corren, D., & Franceschi, J. (1982) Kinetic hydro energy conversion systems study for the New York State resource – Phase I. Final Report. New York Power Authority.

Miller, G., Corren, D., & Franceschi, J. (1985) Kinetic hydro energy conversion systems study for the New York State resource – Phase II and Phase III model testing. *Waterpower'85. Conf. Hydropower*, Las Vegas, NV, 12.

Miller, G., Corren, D., Franceschi, J., & Armstrong, P. (1983) Kinetic hydro energy conversion systems study for the New York State resource – Phase II. Final Report. New York Power Authority.

NOAA (2008) http://tidesandcurrents.noaa.gov/index.shtml [accessed 20 August 2008].

Panmure Gordon (2006) Renewable Energy Holdings Ltd Equity Research, 9 November 2006. Panmure Gordon (UK) Ltd, 28 pp. plus Appendices.

Pugh, D.T. (2004) *Changing Sea Levels: Effects of Tides, Weather and Climate*. Cambridge University Press, Cambridge, UK, 265 pp.

Soulsby, R.L. (1983) The bottom boundary layer of shelf seas, in *Physical Oceanography of Coastal and Shelf Seas*, ed. by Johns, B. Elsevier, Amsterdam, The Netherlands, Chapter 5.

Sternberg, R.W. (1966) Boundary layer observations in a tidal current. *J. Geophys. Res.* **71**, 2175–2178.

UKHO (2002) *Admiralty Sailing Directions West Coasts of England and Wales Pilot*, 15th edition, 383 pp.

UKHO (2003) *Admiralty Sailing Directions North Sea (West) Pilot*, 245 pp.

van Veen, J. (1938) Water movements in the straits of Dover. *J. du Conseil*. **13**, 7–36.

Warhaft, Z., 1997. *An Introduction to Thermal-fluid Engineering: The Engine and the Atmosphere*. Cambridge University Press, UK. 241 pp.

3

Principles of tidal power devices

3.1 Introduction

This chapter begins with a presentation of tidal power devices as a series of non-linear processes, each represented by a series of state variable inputs and outputs. Thus, the hydraulic, turbine, and electrical processes are considered, and equations are derived for the transfer functions. The various tidal stream power devices are then considered in the context of these principles, and the advantages and disadvantages of the different approaches are critically compared.

The Analysis of Tidal Stream Power   Jack Hardisty
© 2009 John Wiley & Sons, Ltd

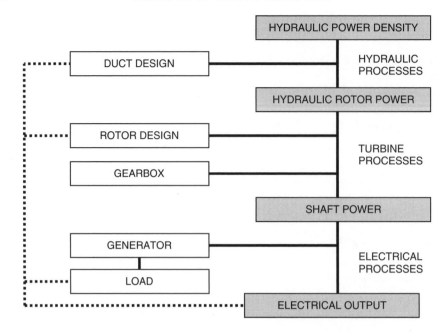

Figure 3.1 Systems diagram for a turbine-based, tidal stream power device, showing the four state variables and the three transfer functions described in the text.

A description of tidal stream power devices is shown in Figure 3.1. There are three sets of transfer functions:

1. **Hydraulic processes.** Local conditions provide the independent control which is the hydraulic power density. The hydraulic power density will vary through the boundary layer profile and throughout the tidal cycle, and through the Spring–Neap and semi-annual cycles, as described in the preceding chapter. The hydraulic process generates an incident current speed at the rotor or turbine that depends upon the form of any duct and the operation of any deflector plates or throttle that may be incorporated into the design.

2. **Turbine processes.** The turbine generates lift or drag on the blades in order to convert the kinetic energy of the tidal current into a circular motion, with a rotation rate and shaft torque that depend upon the design, and with such controlled or fixed variables as the orientation and number of the blades. The turbine generates an output shaft power that may incorporate one or more gearboxes.

3. **Electrical processes.** The generator will then produce an electrical output that will also be controlled by varying the electrical load and will probably be transformed to provide a synchronous grid voltage and frequency.

3.2 Tidal wheels, mills, and barrages

Traditionally, tidal power generation has been associated with the building of a dam or barrage at a coastal site and the generation of power by low-head hydroturbines either during the ebb or during the flood or both. Because of the civil construction costs involved, and the potential impact on the environment, and in spite of there being several candidate sites in existence, no large plant has yet been built in the UK. In comparison with tidal barrages, the direct use of tidal currents avoids large-scale civil engineering activity, and the environmental impacts are lower. Therefore, the subject of this book is direct tidal stream power. However, the work would be incomplete without some consideration of tidal barrages and their history.

A tide mill is a specialist type of water mill driven by tidal rise and fall. A dam or barrage with a sluice is created across a suitable tidal inlet, or a section of river estuary is made into a reservoir. As the tide comes in, it enters the mill pond through a one-way gate, and this gate closes automatically when the tide begins to fall. When the tide is low enough, the stored water can be released to turn a water wheel (Figure 3.2).

The use of tidal wheels, mills, and barrages can be dated back into antiquity. Tide mills were once an important part of the economy of many countries, such as Great Britain and the United States of America – the latter having many hundreds of tide mills operating on the eastern coast from the seventeenth to the nineteenth centuries. Tidal power was harnessed in this fashion not only for milling flour, but for everything from sawing lumber and operating the bellows and hammers in ironworks, manufacturing paper and cotton, to grinding spices, pepper, and gunpowder.

The earliest excavated tide mill, dating from AD 787, is the Nendrum Monastery mill on an island in Strangford Lough in Northern Ireland. ETMTL (2007) describes a mill at Eling on Southampton Water on the English Channel that is more than 900 years old. Eling Tide Mill, although abandoned in the 1940s, survived until it was restored between 1975 and 1980, at which time it reopened as both a working mill and a museum. The earliest surviving reference to it is in the Doomsday Book in AD 1086. Housemill (2007) reports on two tide mills on the River Lea, off the Thames in East London, and Woodbridge (2007) describes a twelfth-century mill that was taken over by a preservation trust in the 1970s and restored. At one time there were 750 tide mills operating on the Atlantic coast of the USA (Minchinton, 1979).

Interest was revived more recently when tidal barrages were built at St Malo in the English Channel (Frau, 1993) and at Kislogubskaya on the Arctic coast of the

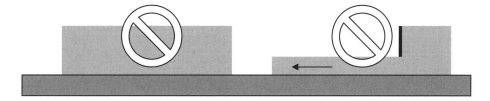

Figure 3.2 The operation of the tidal mill: left, the rising tide floods the basin; right, the falling tide leaves the basin through a sluice driving the water wheel.

Figure 3.3 Tide mills at (left) Olhão, Portugal, and (right) l'Île de Bréhat, France.

Barents Sea (Usachev, Shpolyanskii, & Istorik, 2004). In the UK there are presently proposals to build a tidal barrage at sites in the Severn Estuary.

3.3 Efficiency, capacity factor, and solidity ratio

3.3.1 Efficiency

There are three power efficiency factors that can be combined into an overall device efficiency:

(a) The **hydraulic efficiency**, e_H, is the ratio of the hydraulic power at the rotor or turbine, P_R, to the incident hydraulic power across the capture area of the device, P_H:

$$e_H = \frac{P_R}{P_H} \tag{3.1}$$

(b) The turbine or **rotor efficiency**, e_R, is the ratio of the shaft power, P_S, to the hydraulic power at the rotor or turbine, P_R:

$$e_R = \frac{P_S}{P_R} \tag{3.2}$$

(c) The **electrical efficiency**, e_E, is the ratio of the electrical output power, P_E, to the shaft power, P_S:

$$e_E = \frac{P_E}{P_S} \tag{3.3}$$

(d) The **overall efficiency**, e, is the ratio of the electrical output power, P_E, to the incident hydraulic power, P_H, which from equations (3.1) to (3.3) is

$$e = e_H e_R e_E = \frac{P_R}{P_H}\frac{P_S}{P_R}\frac{P_E}{P_S} = \frac{P_E}{P_H} \tag{3.4}$$

(e) The separate and pragmatic **hydraulic flow efficiency**, e_{HY}, is the ratio of the flow speed at the rotor blade, U_R, to the incident flow speed, U:

$$e_{HY} = \frac{U_R}{U} \qquad (3.5)$$

3.3.2 Capacity factor

The power capacity factor, C_P, is defined as the ratio of the power output over a year to the power that would be output if running at the full, installed capacity, P_I, for the year. There are 8766 hours in an average year, so that

$$C_P = \frac{\int P_E \, dt}{8766 P_I} \qquad (3.6)$$

3.3.3 Solidity ratio

The term **solidity** describes the fraction of the swept area that is solid and will depend upon the number of blades and the width of each blade. Turbines such as the OpenHydro device have high solidity ratios, while those such as Marine Current Turbine's SeaGen have low solidity ratios.

The purpose of this chapter is to derive equations for the hydraulic power at the turbine, the shaft power, and the electrical power, and hence to derive mathematical expressions for the different types of tidal stream power device.

3.4 Hydraulic processes, Part I: Tidal stream power

The **tidal stream power density** is the kinetic energy of the fluid that passes through a unit area of the flow normal to the dominant flow direction. We saw, in Chapter 2, that the tidal hydraulic power density is related to the cube of the flow velocity:

$$P_D = \frac{1}{2}\rho U^3 \qquad (3.7)$$

The density of sea water decreases with an increase in temperature and increases with an increase in salinity. A typical value is about $\rho = 1026 \, \text{kg m}^{-3}$ for a temperature of 20 °C and a salinity of 35 ‰ (gm kg^{-1}).

The tidal stream power devices intersect the flow over a capture area, A_o (m^2), so that the potential hydraulic power incident on each device is

$$P_H = \frac{1}{2}\rho A_o U^3 \qquad (3.8)$$

A typical tidal stream running at 2, 4, 6, and 8 knots (1.02, 2.04, 3.06, and 4.08 m s^{-1}) thus has a hydraulic power density of 0.5, 4.4, 14.7, and 34.8 kW m^{-2}. Alternatively, for an underwater tidal turbine with a blade radius of 10 m, and thus a capture area of $314\,m^2$, the device's potential hydraulic power is $4.6\,MW$ at a speed of 6 knots.

Estimates of the potential power of a tidal stream device depend upon an assessment of the tidal-current velocity which, as demonstrated in Chapter 2, varies considerably from place to place, so this important estimate is by no means a trivial task. Although tidal *levels* can be predicted by summing up a number of *harmonic constituents*, each of which corresponds to a particular astronomical influence with its own characteristic frequency, the prediction of tidal currents is made more complex by their sensitivity to bathymetry and landmasses. There are only three sources of tidal-current information that can be reliably used for the assessment of the hydraulic power at a specific site:

1. **Hydrographic chart data.** The British Admiralty charts may contain current meter data from measurements made in the past for the site of interest. Similar data are available from other charts, as discussed in later chapters. These data are of four, increasingly sophisticated, forms:

   - **Charted maximum current arrows.** Some Admiralty charts with a resolution equal to or better than 1:200 000 contain *tidal arrows* indicating the direction and magnitude of the Spring tide currents, sometimes for both the flood and the ebb tide.
   - **Tidal stream atlases** published by the United Kingdom Hydrographic Office (UKHO) cover specific areas in more detail. Each atlas contains 13 hourly charts with spot values for Spring and Neap tidal velocities, and arrows to indicate the general direction of the currents. Scaling of the arrows may contain further information on the velocity. Each of the 13 charts in any series represents a time in hourly intervals from 6 hours before to 6 hours after high water at Dover.
   - **Tidal diamonds** contain more information and are a feature on some Admiralty charts. For a number of selected points that are marked with diamond symbols, a table lists the hourly velocities up to 6 hours before and 6 hours after high water at a reference port, for average Spring and for average Neap tides. The values shown on the charts are obtained from short-term measurements and are applicable 5 metres below the surface.

   The 'TotalTide' software package was designed by the UKHO for mariners and allows prediction of tidal currents at tidal diamond locations at any time in the past or in the future.

2. **Computational fluid dynamics.** When tidal currents are needed for an area where no measurements have been made, a complete *computational fluid dynamics* (CFD) analysis could be set up. Because of the amount of data required adequately to describe the bathymetry of sea areas and the shapes of landmasses, such studies are usually confined to smaller areas of a few hundred square kilometres. The data in the UK Marine Renewable Energy Atlas (DTI, 2004) is based on interleaved computer models with different resolutions. Such models perform well when given accurate boundary conditions, although at inshore locations the level of accuracy is generally lower.

3. **Current metering.** Definitive information on current velocities, directions, and depth profiles at any location, at a particular site, are usually only available with direct on-site measurements. These should be carried out over at least a complete Spring and Neap cycle. The results should be corrected for the effects of weather.

3.5 Hydraulic processes, Part II: Ducts and diffusers

It will be apparent that some of the devices described in Chapter 4 utilize ducts in order to accelerate the flow onto the turbine and to diffuse the outgoing flow, which can provide a pressure drop and increase the efficiency of the device. In practice, of course, since the ducting is essentially symmetrical so that the device can operate without reorientation in the reversing, tidal flows, the duct acts as both accelerator and diffuser. Two examples, from Lunar Energy and from Clean Current, are shown in Figure 3.4.

The following analysis is based upon Advantage (2007a, 2007b, 2007c) and is presented here in the original notation and then in the notation of the rest of this book as equations (3.17) and (3.19).

Consider a tidal flow that enters a duct at section 1, converges to the upstream face of the turbine at section 2, exits the turbine at section 3 and finally diverges (diffuses) at section 4. The inlet and outlet areas are taken as A_{in} and A_{out}. The power generated by the turbine is the product of the energy extracted from the flow (i.e. the loss in total pressure in the Bernoulli equation) and the volume flowrate, V' ($m^3 s^{-1}$):

$$Power = (P_2 - P_1)\, V' \qquad\qquad (3.9)$$

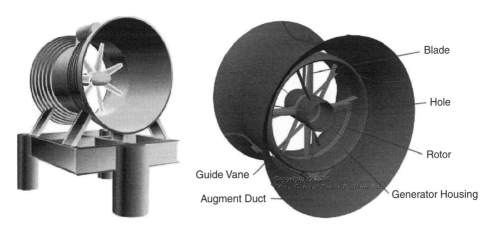

Figure 3.4 RTT concept from Lunar Energy (2008) (left) and the shrouded rotor from Clean Current (2008) (right).

Applying Bernoulli from the inlet to the turbine (section 1 to section 2) gives

$$P_1 = P_2 = P_{S\infty} + \frac{1}{2}\rho\,(v_\infty)^2 \tag{3.10}$$

where $P_{S\infty}$ is the freestream static pressure and v_∞ is the freestream flow velocity. Working in gauge pressure (i.e. $P_{S\infty} = 0$), then

$$P_2 = \frac{1}{2}\rho\,(v_\infty)^2 \tag{3.11}$$

Applying the Bernoulli equation from the turbine to the outlet (section 3 to section 4) gives

$$P_3 = P_4 = P_{S4} + \frac{1}{2}\rho\,(v_4)^2 \tag{3.12}$$

Expressing the exit velocity in terms of the flowrate yields

$$v_4 = \frac{V'}{A_{\text{out}}} \tag{3.13}$$

Therefore, by combining equations (3.12) and (3.13), we obtain

$$P_3 = P_{S4} + \frac{1}{2}\rho\left(\frac{V'}{A_{\text{out}}}\right)^2 \tag{3.14}$$

At the exit, the static pressure equals the freestream pressure:

$$P_3 = \frac{1}{2}\rho\left(\frac{V'}{A_{\text{out}}}\right)^2 \tag{3.15}$$

Substituting equations (3.11) and (3.15) into equation (3.9) yields

$$Power = V'\left[\frac{1}{2}\rho\,(v_\infty)^2 - \frac{1}{2}\rho\left(\frac{V'}{A_{\text{out}}}\right)^2\right] \tag{3.16}$$

Thus, the power depends upon the volume flowrate, V', more commonly called the discharge through the device, Q, upon the freestream velocity, v_∞ (m s^{-1}), and upon the exit area, A_{out} (m^2). Tidal ducts are generally symmetrical, so that $A_{\text{out}} = A$, the simple capture (and exit) area. Equation (3.16) can be rewritten with a value of $\rho = 1000\,\text{kg m}^{-3}$ as

$$Power = \frac{Q}{2}\left[v_\infty^2 - \left(\frac{Q}{A}\right)^2\right]\text{kW} \tag{3.17}$$

Differentiating equation (3.16) with respect to the volume flowrate gives

$$\frac{d\,(Power)}{dV'} = \frac{1}{2}\rho\,(v_\infty)^2 - \frac{3}{2}\rho\left(\frac{V'}{A_{out}}\right)^2 \tag{3.18}$$

Replacing $V' = Q$, the maximum power occurs when equation (3.18) equals zero:

$$Q_{max\,power} = \frac{1}{\sqrt{3}}v_\infty A \tag{3.19}$$

The results are plotted in Figure 3.5 and show that the potential duct power increases with the freestream velocity and the capture/exit area, but maximizes for a particular discharge. For example, a capture area of $100\,m^2$ has the potential to generate $>500\,kW$ at $3.0\,m\,s^{-1}$ and a discharge of about $175\,m^3\,s^{-1}$, which is in agreement with equation (3.19) for $Q \approx 0.6 \times 3.0 \times 100 = 180\,m^3\,s^{-1}$.

Acceleration and diffusion were tested with a 1/10 scale model of the Mark I Neptune Proteus vertical rotor turbine shown in Figure 3.6. The Mark I turbine has a duct consisting of an internal deflector plate set at 45°, although this aligns with the stream on reversal and the internal angles are less severe. The device was tested in the University of Hull flume, where its overall width of 0.70 m was only about one-third of the width of the channel, so that the flow was less constrained to enter through the device. Typical results for the hydraulic efficiency of the device are shown in Figure 3.7.

The results suggest that the hydraulic efficiency of the device increases with both the Froude and the Reynolds numbers (Chapter 2) and can more than double the flow speeds in the device.

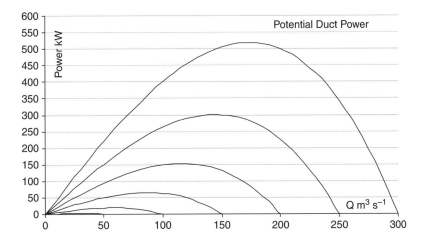

Figure 3.5 Potential duct power from equation (3.17) for a capture area of $100\,m^2$, a range of discharges, and for freestream velocities of 0.5–$3.0\,m\,s^{-1}$ in increments of $0.5\,m\,s^{-1}$ (lower to upper plots). Data from the model on the book's Internet site.

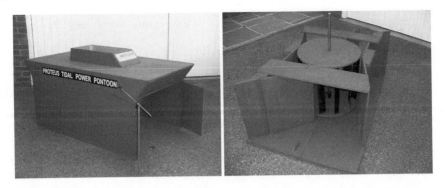

Figure 3.6 The 1/10 scale model of the Mark I Neptune Proteus vertical-axis tidal turbine (left), and a full view with the original single pontoon (right).

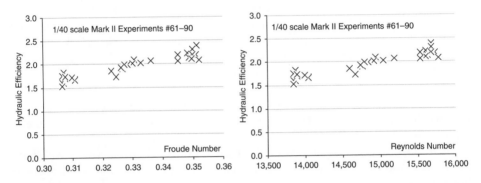

Figure 3.7 Results for the hydraulic efficiency of the 1/40 scale model of the Mark I Proteus vertical rotor tidal turbine described in the text.

3.6 Turbine processes, Part I: Drag and lift forces

3.6.1 General principles

Tidal turbines convert the flow power into rotary motion in order to drive a generator and produce electricity. Consider, firstly, the most common type of blade consisting of an aerofoil (or hydrofoil) section as shown in Figure 3.8, which is effectively a variant of the Kaplan or propeller turbine discussed in Chapter 1.

The **angle of attack** is the angle that an object makes with the direction of the water flow, measured against a reference line in the object and signified by α.

The **chord line** is the reference line from which measurements are made.

In both horizontal- and vertical-axis devices, the resultant force on the turbine blades, resolved into a direction perpendicular to the radius, is responsible for turning the rotor. The water flow over the convex upper surface is accelerated so that, according to Bernoulli's theorem (Chapter 2), a large reduction in pressure is generated, known as the **lift force** on the blade. There is also a reduction in pressure

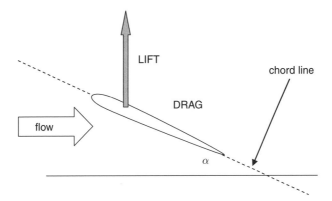

Figure 3.8 Definition diagram for the forces on a blade, as described in the text.

in a downstream direction because of the interference of the blade with the flow, resulting in a **drag force**.

The actual flow velocity experienced at the blade, U', is the difference between the tidal velocity (resolved in a direction parallel to the chord line), U_R, and the blade velocity, V:

$$U' = (U_R - V) \tag{3.20}$$

Albert Betz showed, in 1928, that the maximum fraction of the flow energy that can theoretically be extracted by the turbine is 16/27 (59.3 %) (Taylor, 2004). This occurs when the undisturbed wind velocity is reduced by one-third. The value of 59.3 % is often referred to as the **Betz limit**, and it is believed to apply to both vertical- and horizontal-axis turbines.

The tidal current exerts both drag and lift forces on the turbine, depending upon the shape and orientation of the turbine blades, and these forces exert a rotary torque onto the shaft which then drives electrical generation plant through fixed or variable gearboxes. The principal properties of the drag and lift processes are described in the following section.

3.6.2 Drag forces

Consider water flowing around an immersed object such as the plate, tube, and hydrofoil in Figure 3.9. At a certain point, known as the separation point, the boundary layer becomes detached from the surface and the vorticity is discharged into the body of the fluid (Andrews & Jelley, 2007). The vorticity is transported downstream in the wake. Thus, the pressure distribution on the upstream side and the downstream side are not symmetrical, and the body experiences a net force known as the drag force:

> The **drag force** is the force exerted on an object by fluid flow, acting in the direction of the fluid flow.

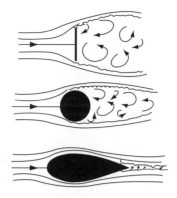

Figure 3.9 Fluid flow around (upper to lower) a flat plate ($C_D = 2.0$), a cylinder ($C_D = 1.2$) and a streamlined section ($C_D = 0.12$) for Reynolds numbers of about 10^5 (after NASA, 2008).

An algebraic form of the drag force can be derived by dimensional analysis as

$$F_D = \tfrac{1}{2}C_D\rho(U_R - V)^2 A \qquad (3.21)$$

Equation (3.21) has the same form as the bed shear stress given in Chapter 2 and illustrates the generic principle that the force increases with the square of the flow speed. The drag coefficient depends upon the shape of the object and upon the Reynolds number of the flow, as indicated by Figure 3.10.

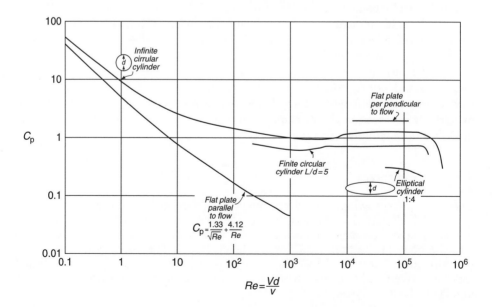

Figure 3.10 Drag as a function of the Reynolds number (after Doppler Systems, 2001).

3.6.3   Lift forces

In a similar manner, the **lift force** is exerted on an object by fluid flow, acting in the direction perpendicular to the fluid flow:

$$F_L = \tfrac{1}{2} C_L \rho (U_R - V)^2 A \qquad\qquad (3.22)$$

where C_L is the lift coefficient. The lift and drag coefficients also depend upon the angle of attack, as shown in Figure 3.11.

The lift coefficient increases linearly to a maximum of about 1.6 at an angle of attack of about 18°, and then, as the hydroplane stalls, the lift drops dramatically. The drag coefficient also increases, but less rapidly, and with an exponential form, achieving a value of about 0.4 at an angle of attack of about 27° and increasing to more than 1 for perpendicular flat blades, and to more than 2 for the open faces of half-pipes mounted perpendicular to the flow. Note that Figure 3.11 applies to

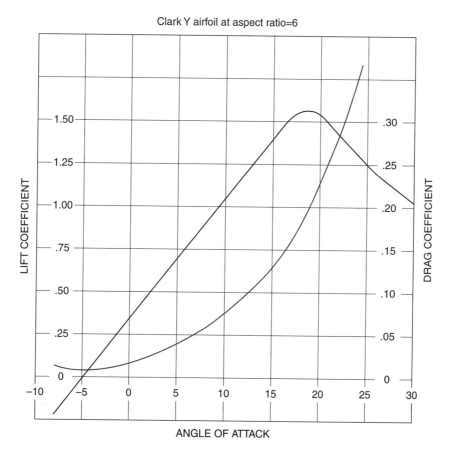

Figure 3.11   Lift (upper line at 10°) and drag (lower line at 10°) coefficients as functions of the angle of attack of the blade (after Search, 2008).

a particular profile, and that the lift coefficient scale is 5 times larger than the drag coefficient scale.

3.7   Turbine processes, Part II: Shaft power

3.7.1   General equations

The turbine or rotor in a tidal power device converts the kinetic energy (and, exceptionally, the potential energy of the flow in ducted turbines, as discussed earlier) into rotary energy in the output shaft. If the rate of revolution of the rotor is n (rev min^{-1}), then the period of rotation $T_S = 60/n$. The tip speed of the rotor blades, V (m s^{-1}), is simply the ratio of the circumference of the rotor to the rotation period $V = 2\pi R_R/T_S$, where R_R is the radius of the rotor.l

The **tip speed ratio**, T_{SR}, is given by the ratio of the blade tip velocity to the fluid velocity:

$$T_{SR} = \frac{V}{U_R} = \frac{2\pi R}{T_S U_R} \tag{3.23}$$

Alternatively, for a given tip speed ratio, the rotational period can be determined from

$$T_S = \frac{2\pi R}{T_{SR} U_R} \tag{3.24}$$

Finally, recalling that the flow speed at the rotor may be given in terms of the independent current speed through the hydraulic flow efficiency (equation (3.5)), we have

$$U_R = e_{HY} U \tag{3.25}$$

Then, the rotor velocity can be given in terms of the independent flow speed by combining equations (3.12) and (3.14) as

$$V = T_{SR} e_{HY} U \tag{3.26}$$

The **torque**, T (N), is the force exerted tangentially on the shaft by the rotor blades and is given by the product of the fluid force, F_F, and the distance of the point of action from the axis of rotation, R_F (m):

$$T = F_F R_F \tag{3.27}$$

The **shaft power**, P_S (W), is then given (e.g. Serway & Jewett, 2004) by

$$P_S = N' \frac{2\pi T}{T_S} \tag{3.28}$$

The **effective blade number**, N', is the number of blades fully impacted by the flow at any one time. As detailed below, a horizontal-axis machine usually has N' equal to the actual number of blades, but this may not be the case in vertical-axis machines.

The shaft power can be determined from a knowledge of the dimensions and rotational speed of the rotor (or the tip speed ratio) and the force on the blades. The force is due to either blade drag or blade lift (which were discussed earlier).

Thus, equation (3.21) or (3.22) (with $C' = C_L$ or $C' = C_D$, as appropriate) can be combined with equations (3.16) and (3.17) to simulate the shaft power:

$$P_S = N'\frac{2\pi T}{T_S} = N'\frac{2\pi F_F R_F}{T_S} = N'\frac{\pi \rho C'(U_R - V)^2 A R_F}{T_S} \quad (3.29)$$

Replacing the turbine period with the tip speed ratio (equation (3.24)) yields

$$P_S = N'\frac{\rho C' T_{SR} U_R (U_R - V)^2 A R_F}{2R} \quad (3.30)$$

Replacing the rotor flow speed with the product of the hydraulic flow efficiency and the freestream velocity (equation (3.25)) yields

$$P_S = 0.5 N' \rho C' T_{SR} e_{HY} U (e_{HY} U - V)^2 A \frac{R_F}{R} \quad (3.31)$$

Replacing V from equation (3.26) yields

$$P_S = 0.5 N' \rho C' T_{SR} e_{HY}^3 U^3 (1 - T_{SR})^2 A \frac{R_F}{R} \quad (3.32)$$

Finally, bring the device specific parameters together into a new, non-dimensional coefficient, J:

$$J = 0.5 N' C' e_{HY}^3 \frac{R_F}{R} \quad (3.33)$$

so that equation (3.21) becomes

$$P_S = J \rho A T_{SR} U^3 (1 - T_{SR})^2 \quad (3.34)$$

Equation (3.21) can be compared with the potential hydraulic power from equation (3.8):

$$P_H = \frac{1}{2} \rho A_o U^3 \quad (3.8)$$

The results are shown in Figure 3.12 for N', C', e_{HY}, R_F, R, A, and ρ of 2.0, 1.6, 1.0, 10 m, 20 m, 30 m², and 1026 kg m⁻³ respectively. In each case the shaft power increases to a maximum at about 0.35, and the simulation here suggests an output

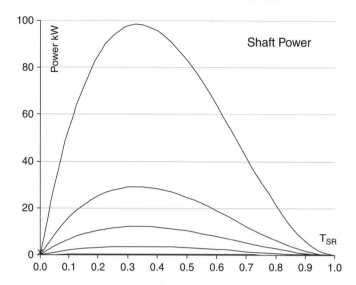

Figure 3.12 Total shaft power (kW) as a function of the tip speed ratio for flow speeds from 0.5 to 3.0 m s^{-1} (lower to upper plot) in increments of 0.5 m s^{-1}, using equations (3.33) and (3.34). From the online model in this book's website.

of about 100 kW at a flow speed of 3.0 m s^{-1}. The tip speed ratio of 0.35 applies to the radius at the mean action of the torque, which in the present case is $R_F/R = 0.5$, so that the 'true' T_{SR} is R^2/R_F^2 larger at about 1.4.

3.7.2 Horizontal-axis devices

The thrust on the turbine in horizontal-axis devices is generated and translated into rotational energy by shaping the turbine blades as hydrofoils. Andrews and Jelley (2007) explain that, since the apparent angle of the flow decreases closer to the hub (because the blade is moving more slowly), turbine blades are designed with a twist that increases closer to the hub so that the angle of attack remains at the optimum. The blade width also increases closer to the hub so that the component of lift generates the lift required to maintain the Betz condition. The result is shown, for example, in Marine Current Turbines' Seaflow Project in the Severn Estuary with variable-width blades that are naturally twisted (Figure 3.13).

Thus, we have the definition of the transfer function for the hydraulic processes in a lift-based tidal turbine from equations (3.33) and (3.34). The apparent number of blades is the same as the actual number of blades, since all blades are impacted by the freestream flow throughout the rotation. The coefficient is replaced by the lift coefficient, and the device has variable-pitch blades to maintain the angle of attack as close to the optimum of about 15–18°. The hydraulic flow efficiency, e_{HY}, remains close to unity, but the flow can be slowed by an effect known as the axial interference factor (Taylor, 2004, p. 262). For the present purposes, we set $e_{HY} = 1$. The ratio R_F/R, which is the ratio of the radius of the point of action of the total torque to the full blade radius, depends upon the shape and twist of the individual

Figure 3.13 Variable-width and twist blades, which also have power pitch controls, as evidenced by the Marine Current Turbine's Seaflow device from Marine Current Turbines (2008).

blades, but usually lies between 0.5 and 0.8. Equation (3.33) for the horizontal-axis coefficient, J_{HA}, thus becomes

$$J_{HA} = 0.5NC_L \frac{R_F}{R} \tag{3.35}$$

Equation (3.34) for the horizontal-axis shaft power, P_{SHA}, thus becomes

$$P_{SHA} = J\rho A T_{SR} U^3 \tag{3.36}$$

where A is the product of the chord length and the blade length. As discussed earlier, T_{SR} is the blade speed at the radius of the point of action of the total torque, which is smaller than the total radius. The $(1 - T_{SR})$ term, which represents the difference between the flow speed and the blade speed, assumes a value of unity, as the blade does not move downstream.

The results are shown in Figure 3.14 for N, C_L, e_{HY}, R_F, R, A, and ρ of 3.0, 1.6, 1.0, 10 m, 20 m, 30 m², and 1026 kg m⁻³ respectively. In each case the shaft power increases with the tip speed ratio, and the simulation here suggests an output of about 1000 kW at a flow speed of 3.0 m s⁻¹ and $T_{SR} = 1.0$. The tip speed ratio applies to the radius at the mean action of the torque, which in the present case is $R_F/R = 0.5$, so that the 'true' T_{SR} is R^2/R_F^2 larger at about 4.0.

Finally, if the tip speed ratio is too large, then the advancing blade enters interference from the preceding blade. Alternatively, if the tip speed ratio is too small, then too much water flows through the turbine without doing useful work on the blades.

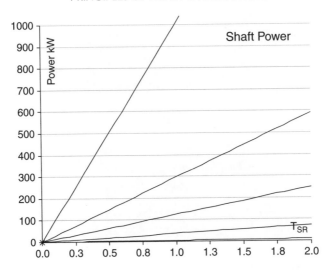

Figure 3.14 Shaft power from a lift turbine with three blades, as described by equation (3.36). Results from the online model on this book's website.

3.7.3 Vertical-axis devices

The thrust on the turbine in vertical-axis devices is generated and translated into rotational energy by utilizing either the drag of the blades (Figure 3.15 (left)) or the lift of the blades (Figure 3.15 (right)). Unlike the horizontal-axis devices described above, vertical-axis device blades can retain a constant width, as the blade velocity

Figure 3.15 (Left) The Neptune Renewable Energy, Mark III, Proteus tidal turbine is a vertical-axis device utilizing drag on the multiple blades. Reproduced with permission from www.neptunerenewableenergy.com. (Right) The Kobold vertical-axis lift blade turbine. Reproduced courtesy of Prof. D.P. Coiro, ADAG Research Group, Dept of Aerospace Engineering, University of Naples 'Federico II', Italy.

does not change along the length of the blade. In the present section we focus on vertical-axis, drag-dominated applications.

We can again define the transfer function for the hydraulic processes in such a drag-based tidal turbine from equations (3.33) and (3.34). The apparent number of blades is less than the actual number of blades, since blades are not impacted by the freestream flow throughout the full rotation. The coefficient is replaced by the drag coefficient (and lift blades again have variable-pitch blades to maintain the angle of attack as close to the optimum of about 15–18°. The hydraulic flow efficiency, e_{HY}, may be less than unity, but can be enhanced by appropriate ducting as described above. The ratio R_F/R, which is the ratio of the radius of the point of action of the total torque to the full blade radius, is much closer to unity than for the horizontal-axis turbines described above. Equation (3.33) for the vertical-axis coefficient, J_{VA}, thus becomes

$$J_{VA} = 0.5N'C_D e_{HY}^3 \frac{R_F}{R} \tag{3.37}$$

Equation (3.34) for the horizontal-axis shaft power, P_{SVA}, thus becomes

$$P_{SVA} = J\rho AT_{SR} U^3 (1 - T_{SR})^2 \tag{3.38}$$

where A is the product of the blade width and the blade length. The $(1 - T_{SR})$ term, which represents the difference between the flow speed and the blade speed, is now less than unity, as the blade travels less quickly than the flow.

The results are shown in Figure 3.16 for N', C_D, e_{HY}, R_F, R, A, and ρ of 8.0, 2.0, 1.0, 6 m, 6 m, 12 m^2, and 1026 kg m^{-3} respectively. In each case the shaft power

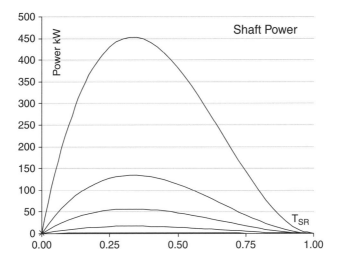

Figure 3.16 Shaft power from a 16-bladed drag vertical-axis turbine, as described by equation (3.38). Results from the online model on this book's website.

increases with the tip speed ratio, and the simulation here suggests an output of about 450 kW at a flow speed of 3.0 m s^{-1} and $T_{SR} = 0.35$.

Finally, if the tip speed ratio is too large, then the advancing blade enters interference from the preceding blade. Alternatively, if the tip speed ratio is too small, then too much water flows through the turbine without doing useful work on the blades.

3.8   Electrical processes, Part I: Overview and generators

3.8.1   Overview

There are two basic approaches to the generation of electrical potential difference from the rotational energy of the shaft. In synchronous systems the magnetic field rotates, whereas in asynchronous generators the conductors rotate.

3.8.2   Synchronous motors

All three-phase generators (or motors) use a rotating magnetic field, as shown in Figure 3.17 (Danish Wind Energy Association, 2008). Three electromagnets are installed around a circle, and each of the three magnets is connected to its own phase in the three-phase electrical grid. The three electromagnets thus alternate between producing a south pole and a north pole towards the centre corresponding to the fluctuation in voltage of each phase. Thus, the magnetic field will make one complete revolution per cycle. The compass needle will follow the magnetic field exactly, and make one revolution per cycle. With a 50 Hz grid, the needle will make 50 rev s^{-1} or 3000 rpm.

The diagram illustrates a two-pole permanent magnet synchronous motor, as it has only one north and one south pole, is synchronous with the mains supply, and the compass needle is a permanent magnet, not an electromagnet. The three electromagnets are called the stator because this part remains static. The compass needle is called the rotor, because it rotates.

3.8.3   Synchronous generators

If the magnet in Figure 3.17 is driven around (instead of letting the current from the grid move it), the system works as a generator, sending alternating current back

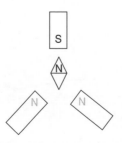

Figure 3.17   Two-pole permanent magnet motor with three coils.

into the grid. The more force (torque) you apply, the more electricity you generate, but the generator will still run at the same speed dictated by the frequency of the electrical grid. If the generator is disconnected from the main grid, however, it will have to be cranked at a constant rotational speed in order to produce alternating current with a constant frequency.

In practice, permanent magnet synchronous generators are rarely used because permanent magnets tend to become demagnetized by working in the powerful magnetic fields inside a generator and because of the high cost of the powerful magnets that are made of rare-earth metals such as neodymium.

3.8.4 Asynchronous (induction) generators

The asynchronous generator has a rotor consisting of a number of copper or aluminium bars which are connected electrically by aluminium end-rings. For obvious reasons this is known as a cage rotor (Figure 3.18). The rotor is provided with an 'iron' core, using a stack of thin insulated steel laminations, with holes punched for the conducting aluminium bars. The rotor is placed in the middle of the stator, which is a four-pole stator directly connected to the three phases of the electrical grid.

3.8.5 Asynchronous motor

The machine will start turning like a motor at a speed that is just slightly below the synchronous speed of the rotating magnetic field from the stator when a current is connected. This is because a magnetic field is generated that moves relative to the rotor, inducing a very strong current in the rotor bars which offer very little resistance to the current, since they are short circuited by the end-rings. The rotor then develops its own magnetic poles, which in turn become dragged along by the electromagnetic force from the rotating magnetic field in the stator.

3.8.6 Asynchronous generator

No current is generated if this arrangement is cranked around at exactly the synchronous speed of the generator described earlier (1500 rpm) because the magnetic field rotates at exactly the same speed as the rotor so that there is no induction

Figure 3.18 Two-pole permanent magnet motor.

phenomena in the rotor, and it will not interact with the stator. If, however, the speed is increased above 1500 rpm, the rotor moves faster than the rotating magnetic field from the stator, which means that, once again, the stator induces a strong current in the rotor. This arrangement is known as an asynchronous generator.

3.8.7   Generator slip

The speed of the asynchronous generator will vary with the turning force (moment, or torque) applied to it. In practice, the difference between the rotational speed at peak power and at idle is very small and is called the generator slip. Thus, a four-pole generator will run idle at 1500 rpm if it is attached to a grid with a 50 Hz current. If the generator is producing at its maximum power, it will be running at 1515 rpm.

It is a very useful mechanical property that the generator will increase or decrease its speed slightly if the torque varies, as there will be less tear and wear on the gearbox. This is one of the most important reasons for using an asynchronous generator rather than a synchronous generator on a renewable energy device that is directly connected to the electrical grid.

3.9   Electrical processes, Part II: Cabling and grid connection

DTI (2007) reports on a number of different configuration options for tidal stream generators and shore connection:

Option 1 The first option was a simple system consisting of a single turbine, an AC generator, and a power converter with a step-up transformer. This arrangement is usually used in near-to-shore low-power projects, and would be suitable for a pilot project or technology demonstrator. It would be a poor choice for a farm, as there is no opportunity to take benefit from the common costs of a farm installation.

Option 2 The second option, involving multiple generators, each with a dedicated power converter, would provide a high level of system flexibility, allowing generators to rotate at different speeds as well as individual generators

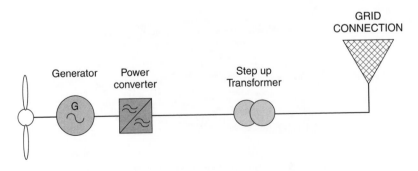

Figure 3.19   Option 1 – single system.

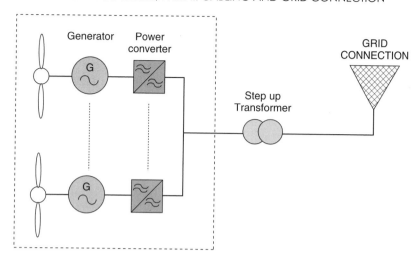

Figure 3.20 Option 2 – multiple turbines, single system.

to be shut down. However, the system is complex, and it would be a high-cost solution with no cable or transformer redundancy.

Option 3 The use of multiple induction generators connected to a single power converter is appropriate where there is little variation in turbine speed across the farm. This is important – high circulating currents between the generators would occur if the differences in speed were significant. In the absence of such large circulating currents, this option offers a lower cost in comparison with the other system configurations for a high-power farm. Again, there is no cable or transformer redundancy, and in this case no inherent converter redundancy is possible without a reserve converter.

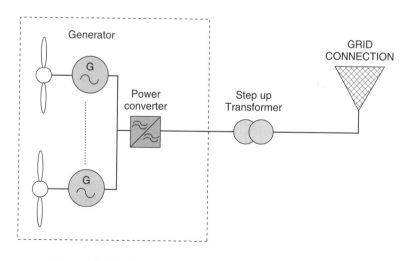

Figure 3.21 Option 3 – multiple induction generators.

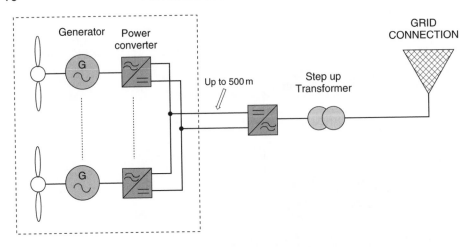

Figure 3.22   Option 4 – active DC link.

Option 4 Under the active DC link option, each generator has a dedicated power converter machine bridge that feeds a common DC link; a single network bridge then feeds the transformer. This option is suitable for projects where generators are located close together but are not necessarily rotating at the same speed. Individual generator control is possible, but, if the converter network bridge needs to be shut down, power generation ceases. The assumed farm capacity of 30 MW would in practice imply the use of two network bridges, each serving half the farm.

3.10   Categories of turbines

Turbines convert the fluid power in water into rotational power in a shaft and are subdivided into **impulse** and **reaction** machines (Douglas, Gasiorek, & Swaffield, 2001).

Impulse turbines convert the total head into kinetic energy through one or more nozzles creating jets that strike vanes attached to the periphery of a rotating wheel. Because of the rate of change in angular momentum and the motion of the vanes, work is done on the runner (impeller) by the fluid, and, thus, energy is transferred. Since the fluid energy which is reduced on passing through the runner is entirely kinetic, it follows that the absolute velocity at outlet is smaller than the absolute velocity at inlet (jet velocity). Furthermore, the fluid pressure is atmospheric throughout, and the relative velocity is constant except for a slight reduction due to friction.

In **reaction turbines**, the fluid passes first through a ring of stationary guide vanes in which only part of the available total head is converted into kinetic energy. The guide vanes discharge directly into the runner along the whole of its periphery, so that the fluid entering the runner has pressure energy as well as kinetic energy. The pressure energy is converted into kinetic energy in the runner, and therefore the relative velocity is not constant but increases through the runner. There is, therefore, a pressure difference across the runner.

The **degree of reaction** describes a measure of the continuum from impulse to reaction turbines. It is derived by the application of Bernoulli's equation to the inlet and outlet of a turbine. Thus, if the conditions at inlet are denoted by the use of suffix 1, and those at outlet by the suffix 2, then

$$\frac{p_1}{\rho g} + \frac{U_1^2}{2g} = E + \frac{p_2}{\rho g} + \frac{U_2^2}{2g} \tag{3.39}$$

where E is the energy transferred by the fluid to the turbine per unit weight of the fluid. Thus

$$E = \frac{p_1 - p_2}{\rho g} + \frac{U_1^2 - U_2^2}{2g} \tag{3.40}$$

In this equation, the first term on the right-hand side represents the drop in static pressure (or potential energy) in the fluid across the turbine, whereas the second term represents the drop in the velocity head (or kinetic energy). The two extreme solutions are obtained by making either of these two terms equal to zero. Thus, if the pressure is constant, so that $p_1 = p_2$, then $E = (U_1^2 - U_2^2)/2g$ and the turbine is purely impulsive. If, on the other hand, $U_1 = U_2$, then $E = (p_1 - p_2)/\rho g$ and the device is a reaction turbine. The continuum is described by the degree of the reaction, R_T, as defined by

$$R_T = \frac{\text{static pressure drop}}{\text{total energy transfer}}$$

The static pressure drop is given by

$$\frac{p_1 - p_2}{\rho g} = E - \frac{U_1^2 - U_2^2}{2g} \tag{3.41}$$

and the total energy transfer is E, so that

$$R_T = 1 - \frac{\left(U_1^2 - U_2^2\right)}{2gE} \tag{3.42}$$

Water turbines are used in power stations to drive electric generators. There are two well-known large-head types (Figure 3.23): the Pelton wheel, which is an impulse turbine, and the Francis type. Table 3.1 compares the two types.

In tidal applications, however, there is a minimal head, and therefore the main application is for the reaction-type crossflow (Ossberger, Savonius, Michell, or Banki) turbine and the axial-flow (Kaplan) turbines. Since most Kaplan-type reaction turbines are **horizontal axis,** they are considered as such below (Section 3.11). Most of the crossflow turbines are **vertical axis,** and are thus considered as such below (Section 3.12). Finally, hydrofoil blades are also utilized in a reciprocating

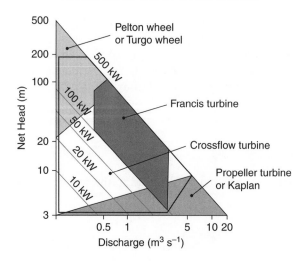

Figure 3.23 The various turbines as a function of the head (vertical axis, m) and the flow discharge (horizontal axis, m³ s⁻¹), after British Hydro, 2008.

Table 3.1 Comparison of water turbines

	Pelton wheel	**Francis**	**Kaplan**
Total head (m)	100–1700	80–500	Up to 400
Maximum power (MW)	55	40	30
Efficiency (per cent)	93	94	94
Regulation mechanism	Nozzle and deflector plate	Guide vanes	Blade stagger

manner or, as in the Kobold and Darius turbines, with crossflow reaction blades, and they are considered as hybrids.

3.11 Horizontal-axis devices

Horizontal-axis turbines (axial-flow turbines) are similar in concept to the horizontal-axis wind turbine (HAWT) and are also a derivative of the Kaplan turbine described earlier. Prototype tidal turbines have been built and tested using this concept, and, in some cases, ducts may be used around the blades to increase the flow and power output from the turbine (Figure 3.24 (left)).

It appears that the British development of horizontal-axis tidal stream power devices can be traced back to a river current turbine project that ran from 1976–1984. One of the instigators was Peter Fraenkel, initially from within ITDG but then in the newly formed IT Power. The turbine used a vertical-axis Darrieus-type rotor, and was moored off the bank of the river Nile in Juba, Sudan, where it was used for irrigation. Following this 'run of river' project, IT Power began work on a turbine designed specifically to produce electricity from tidal currents. The 3.5 m diameter horizontal-axis turbine was suspended beneath a floating raft moored in

Figure 3.24 (Left) Ducted horizontal-axis tidal turbine from OpenHydro under test at the European Marine Energy Centre in the Orkney Islands (OpenHydro, 2008). (Right) Artist's impression of Marine Current Turbines' SeaGen twin-impellor device (Marine Current Turbines, 2008).

the Corran Narrows, at the entrance to Loch Linnhe in Scotland, and produced a shaft power of 15 kW. In the following years, Fraenkel established Marine Current Turbines and went on to design and construct the Seaflow and SeaGen devices (Figures 3.13 and 3.24).

3.12 Vertical-axis devices (crossflow turbines)

Since ancient times there have been many designs of vertical-axis wind turbines (Figure 3.25), most of which have been adopted for tidal stream use. Vertical-axis machines have many operational advantages, largely associated with the fact that all gearing and electrical machinery can be raised above the water line for ease of

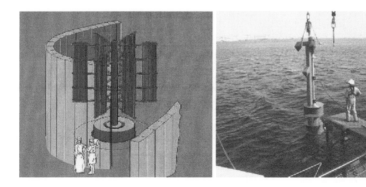

Figure 3.25 (Left) Artist's impression of an ancient vertical-axis windmill. (Right) SeaPower's EXIM tidal Savonius rotor device under test off the north coast of Scotland.

maintenance and repair. There are three basic types of vertical-axis device: firstly the purely drag-based Savonius, secondly the purely lift-based Kobold crossflow and Darrieus rotors (also known as the H-rotor), and thirdly the drag- and impulse-based crossflow (also known as the Ossberger, Michell, or Banki rotor).

1. **Savonius**. The Savonius machine is a vertical-axis machine that uses a rotor invented by Finnish engineer S.J. Savonius in 1922. The approach may be seen as a development of the Savonius geometry described by Charlier (2003) as originating from, for example, Musgrove (1979). In its simplest form it is essentially two cups or half-drums fixed to a central shaft in opposing directions. Each cup catches the flow and so turns the shaft, bringing the opposing cup into the flow. The operation of the device depends upon the difference between the drag coefficients of the open face of the cup, which can be estimated at about 2.5, whereas the drag coefficient on the rear face of the drum is only about 1.2. SeaPower have developed a Savonius rotor tidal device called EXIM, which has been tested off the north coast of Scotland (Figure 3.25 and Chapter 4).

2. **The Kobold and Darrieus crossflow (straight-bladed, spiral-bladed, vertical-axis or H-rotor)**. In 1931, a French engineer, George J.M. Darrieus, invented the Darrieus vertical-axis wind turbine. The turbine was developed as a ducted vertical-axis tidal turbine by Davis and Swan (1982, 1983) and considered for New York's East River by Miller *et al.* (1982, 1983, 1984, 1985). The Darrieus type of machine consists of two or more flexible aerofoil blades, often in spiral forms, which are attached to both the top and the bottom of a rotating vertical shaft. In tidal use, the flow over the hydrofoil contours of the blade creates lift and exerts a turning moment and torque on the shaft.

Figure 3.26 Artist's impression of the Ponte d'Archimede's Kobold H-rotor (from Ponte di Archimede, 2008).

P.J. Musgrove led a research project in 1975 at Reading University to straighten out the blades of a Darrieus wind turbine, so creating a straight-bladed vertical-axis wind turbine or H-rotor configuration. These earlier wind machines with feathering blades were known as variable-geometry vertical-axis wind turbines. A number of marine device developers took up the concept, principally the Blue Energy Company in Canada and the Ponte d'Archimede group in Italy (Figure 3.26). The latter have published some very interesting drag coefficient measurements, as shown in Figure 3.27. The drag coefficient

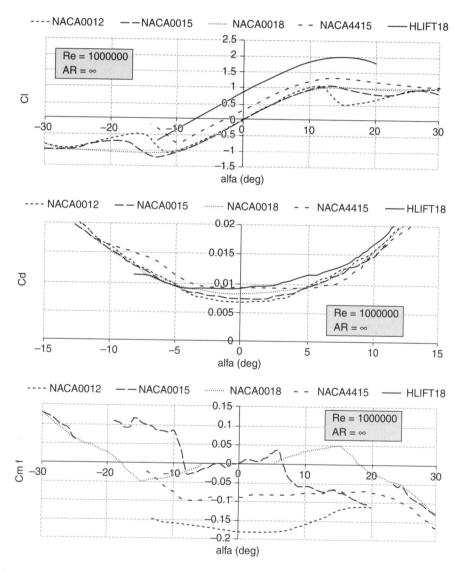

Figure 3.27 Comparison between H-lift and classical airfoil characteristics: top, lift; middle, drag; lower, moment (from University of Naples, 2008).

remains low, but, as described earlier, the lift coefficient increases with the angle of attack, achieving the maximum value of about 2 for an angle of attack of about 16°. The problem resolves into maintaining this optimum as the device's blade rotates, which is achieved by a complex set of levers and cams. The Ponte d'Archimede device was built at full scale and deployed in the Straits of Messina between Italy and Sicily, as described in Chapter 4.

3. **The crossflow (Ossberger, Michell, or Banki) turbine** is a radial freestream turbine. From its specific speed it is classified as a slow-speed turbine (Chapter 1). The guide vanes impart a rectangular cross-section to the water jet. It flows through the blade ring of the cylindrical rotor, first from the outside inward, then, after passing through the inside of the rotor, from the inside outward. This low-head, low-specific-speed crossflow rotor is described in detail by Mockmore and Merryfield (1949) with a vertical axis mounted on a moored pontoon. The crossflow turbine (Ossberger, 2008) has been adopted for use in the tidal environment by Neptune Renewable Energy, as shown in Figures 3.28 and 3.29 and described in Chapter 4. Figure 3.29 (upper) shows the original patent drawings for the Mark III device, and Figure 3.29 (lower) shows the results of computational fluid dynamic (CFD) simulations (Advantage, 2007a, 2007b, 2007c, 2007d).

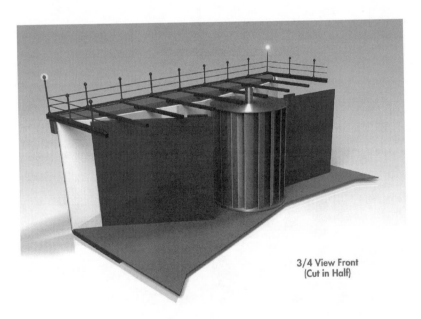

3/4 View Front
(Cut in Half)

Figure 3.28 The Neptune Proteus crossflow tidal turbine (Neptune RE Ltd, 2008).

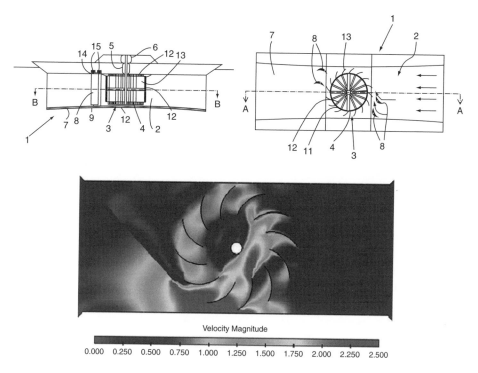

Figure 3.29 (Upper) The original patent drawings for the Neptune Proteus NP1000 crossflow rotor device. (Lower) Results from CFD simulations of this device.

3.13 Summary

The tidal turbine systems may be described as hydraulic, turbine, and electrical processes. The transfer functions may be described by hydraulic, rotor, electrical, overall, and hydraulic flow efficiencies respectively:

$$e_{\mathrm{H}} = \frac{P_{\mathrm{R}}}{P_{\mathrm{H}}} \tag{3.1}$$

$$e_{\mathrm{R}} = \frac{P_{\mathrm{S}}}{P_{\mathrm{R}}} \tag{3.2}$$

$$e_{\mathrm{E}} = \frac{P_{\mathrm{E}}}{P_{\mathrm{S}}} \tag{3.3}$$

$$e = e_{\mathrm{H}}e_{\mathrm{R}}e_{\mathrm{E}} = \frac{P_{\mathrm{E}}}{P_{\mathrm{H}}} \tag{3.4}$$

$$e_{\mathrm{HY}} = \frac{U_{\mathrm{R}}}{U} \tag{3.5}$$

The **capacity factor** C_P is defined as

$$C_P = \frac{\int P_E dt}{8766 P_I} \tag{3.6}$$

The freestream velocity at a tidal site may be synthesized by the harmonic expansion derived in the preceding chapter, and the potential hydraulic power is then given by

$$P_H = \frac{1}{2} \rho A_o U^3 \tag{3.8}$$

The potential hydraulic power within a turbine duct is given by

$$Power = \frac{Q}{2} \left[v_\infty^2 - \left(\frac{Q}{A} \right)^2 \right] \text{kW} \tag{3.17}$$

with a maximum discharge

$$Q_{\text{max power}} = \frac{1}{\sqrt{3}} v_\infty A \tag{3.19}$$

The **drag force** is the force exerted on an object by fluid flow, acting in the direction of the fluid flow:

$$F_D = \tfrac{1}{2} C_D \rho (U_R - V)^2 A \tag{3.21}$$

The **lift force** is the force exerted on an object by fluid flow, acting in the direction perpendicular to the fluid flow:

$$F_L = \tfrac{1}{2} C_L \rho (U_R - V)^2 A \tag{3.22}$$

The **tip speed ratio**, T_{SR}, is

$$T_{SR} = \frac{V}{U_R} = \frac{2\pi R}{T_S U_R} \tag{3.23}$$

The **rotational period**, T_S, is

$$T_S = \frac{2\pi R}{T_{SR} U_R} \tag{3.24}$$

The flow speed at the rotor may be given in terms of the independent current speed through the hydraulic flow efficiency:

$$U_R = e_{HY} U \tag{3.25}$$

and

$$V = T_{SR}e_{HY}U \tag{3.26}$$

The **torque**, $T(\text{N})$, is

$$T = F_{F}R_{F} \tag{3.27}$$

The **shaft power**, P_{S} (W), is

$$P_{S} = N'\frac{2\pi T}{T_{S}} \tag{3.28}$$

Combining the equations yields

$$P_{S} = J\rho AT_{SR}U^{3}(1 - T_{SR})^{2} \tag{3.34}$$

where

$$J = 0.5N'C'e_{HY}^{3}\frac{R_{F}}{R} \tag{3.33}$$

Equations (3.33) and (3.34) can be applied to horizontal-axis lift turbines in the forms

$$J_{HA} = 0.5NC_{L}\frac{R_{F}}{R} \tag{3.35}$$

$$P_{SHA} = J\rho AT_{SR}U^{3} \tag{3.36}$$

Equations (3.33) and (3.34) can be applied to vertical-axis drag turbines in the forms

$$J_{VA} = 0.5N'C_{D}e_{HY}^{3}\frac{R_{F}}{R} \tag{3.37}$$

$$P_{SVA} = J\rho AT_{SR}U^{3}(1 - T_{SR})^{2} \tag{3.38}$$

3.14 Bibliography

Advantage (2007a) *Initial Analysis of the Neptune Tidal Power Pontoon.* TM1354-NE1. Advantage CFD.

Advantage (2007b) *Analysis of Alternative Blade Cionfigurations.* TM1380-NE1. Advantage CFD.

Advantage (2007c) *Analysis of Two Alternative Pontoon Designs.* TM1382-NE1. Advantage CFD.

Advantage (2007d) *Effect of Shroud Extensions and an Inlet Side Expansion.* TM1396-NE1. Advantage CFD.

Andrews, J. & Jelley, N. (2007) *Energy Science: Principles, Technologies and Impacts.* OUP, Oxford, UK, 328 pp.

British Hydro (2008) http://www.british-hydro.co.uk/infopage.asp?infoid=362 [accessed 6 September 2008].

Charlier, R.H. (2003) A sleeper awakes: tidal current power. *Renewable and Sustainable Energy Reviews* 7, 515–529.

Clean Current (2008) http://www.cleancurrent.com

Danish Wind Energy Association (2008) http://www.windpower.org/en/tour/wtrb/electric.htm [accessed 26 March 2008].

Davis, B.V. & Swan, D.H. (1982) Extracting energy from river and tidal currents using open and ducted vertical axis turbines. Model tests and prototype design, in *Proceedings of International Conference on 'New Approaches to Tidal Power'*, Bedford Institute of Oceanography, Dartmouth, NS, Canada.

Davis, B.V. & Swan, D.H. (1983) *Vertical Axis Turbine Economics for Rivers and Estuaries in Modern Power Systems*. Nova Energy Ltd, Montreal, Canada.

Doppler Systems (2001) *Wind Loading on Fixed Site Doppler Antennas*. Technical Application Note. http://www.dopsys.com/loads.htm [accessed 20 August 2008].

Douglas, J.F., Gasiorek J.M. and Swaffield, J.A. (2001). *Fluid Mechanics, 4th Edition*. Pearson, Harlow, Essex. 911pp.

DTI (2004) *Atlas of UK Marine Renewable Energy Resources*. Technical Report.http://www.dti.gov.uk/energy/sources/renewables/renewables-explained/wind-energy/page27403.html [accessed 7 May 2007].

DTI (2007) *Economic Viability of a Simple Tidal Stream Energy Capture Device*. http://www.berr.gov.uk/files/file37093.pdf [accessed 26 March 2008].

ETMTL (2007) http://www.elingtidemill.wanadoo.co.uk/sitem.html [accessed 8 May 2007].

Frau, J.P. (1993) Tidal energy: promising projects: La Rance, a successful industrial-scale experiment. *Energy Conversion* 8, 552–558.

Housemill (2007) http://www.housemill.org.uk/housemill.htm [accessed 8 May 2007].

Lunar Energy (2008) http://www.lunarenergy.co.uk [accessed 20 August 2008].

Marine Current Turbines (2008) http://www.dti.gov.uk/files/file29969.pdf [Accessed 20 August 2008].

Miller, G., Corren, D., & Armstrong, P. (1984) Kinetic hydro energy conversion systems study for the New York State resource – Phase II and Phase III model testing. New York Power Authority.

Miller, G., Corren, D., & Franceschi, J. (1982) Kinetic hydro energy conversion systems study for the New York State resource – Phase I. Final Report. New York Power Authority.

Miller, G., Corren, D., & Franceschi, J. (1985) Kinetic hydro energy conversion systems study for the New York State resource – Phase II and Phase III model testing. *Waterpower'85. Conf. Hydropower*, Las Vegas, NV, 12.

Miller, G., Corren, D., Franceschi, J., & Armstrong, P. (1983) Kinetic hydro energy conversion systems study for the New York State resource – Phase II. Final Report. New York Power Authority.

Mockmore, C.A. & Merryfield, F. (1949) *The Banki Water Turbine*. Oregon State University, Civil Engineering, Department Engineering Bulletin Number 25. http://www.cd3wd.com/CD3WD_40/VITA/BANKITUR/EN/BANKITUR.HTM [accessed 16 January 2008].

Musgrove, P. (1979) Tidal and river current energy systems, in *Proceedings of Conference on 'Future Energy Concepts'*, Institution of Electrical Engineers, London, UK, Vol. 17, pp. 114–117.

NASA (2008) http://history.nasa.gov/SP-367/f38.htm [accessed 30 June 2008].

Neptune RE Ltd (2008) www.neptunerenewableenergy.com [accessed 20 August 2008].

OpenHydro (2008) http://www.openhydro.com/home.html [accessed 20 August 2008].

Ossberger (2008) http://www.ossberger.de/index.php?sprache=en& pageid=1–3 [accessed 30 June 2008].

Ponte di Archimede (2008) Ponte di Archimede (2007) [accessed 18 March 2008].

Search (2008) http://www.search.com/reference/Airfoil [accessed 20 August 2008].

Serway, R.A. and Jewett, J.W. (2004). *Physics for Scientists and Engineers*, Brooks/Cole.

Taylor, D. (2004) Wind energy, in *Renewable Energy: Power for a Sustainable Future*, ed. by Boyle, G. Oxford University Press, Oxford, UK.

University of Naples (2008) http://www.dpa.unina.it/adag/eng/renewable_energy.html [accessed 20 August 2008].

Usachev, I.N., Shpolyanskii, Yu. B., & Istorik, B.P. (2004) Performance and control of a marine power plant in the Russian Arctic coast and prospects for the wide-scale use of tidal energy. *Power Technology and Engineering* 38, 188–206.

4

Tidal power technologies

4.1 Introduction

This chapter introduces 40 technologies that have been proposed for the capture of electrical energy from tidal streams. The range of technologies described by developers' websites, by Bedard (2005), by Pure Energy Systems (2007), and by Moore (2007) are listed in alphabetical order in Table 4.1, along with section

Table 4.1   Alphabetical listing of the devices discussed in this chapter, together with Internet references and section references. A digital copy of the table and the Internet links is available through this book's website

1	Atlantisstrom	http://www.atlantisstrom.de/index_english.html	4.18
2	Aquantis	http://www.iere.org/documents/ tidal.pdf	4.18
3	BioPower Systems	http://www.biopowersystems.com	4.17
4	Blue Energy	http://bluenergy.com/	4.2
5	Bourne Energy	http://bourneenergy.com/	4.17
6	Clean Current	http://www.cleancurrent.com	4.3
7	C-Power	http://www.c-power.co.uk/	4.18
8	Edinburgh POLO	http://www.mech.ed.ac.uk/research/ wavepower/	4.17
9	Engineering Business	http://www.engb.com/services_09a.php	4.4
10	EvoPod	http://www.oceanflowenergy.com/	4.18
11	Gorlov Helical Turbine	http://www.gcktechnology.com	4.5
12	Hammerfest Strøm	http://www.e-tidevannsenergi.com/	4.6
13	Hydrohelix	http://www.hydrohelix.fr	4.17
14	Hydroventuri	http://www.hydroventuri.com	4.18
15	Lunar Energy	http://www.lunarenergy.co.uk/	4.7
16	Marine Current Turbines	http://www.marineturbines.com	4.8
17	Neptune Proteus	http://www.neptunerenewableenergy.com	4.17
18	OpenHydro	http://www.openhydro.com	4.10
19	ORPC OcGen	http://www.oceanrenewablepower.com/	4.18
20	Ponte di Archimede Kobold	Ponte di Archimede (2007)	4.11
21	Pulse Generation	http://www.pulsegeneration.co.uk/	4.12
22	ScotRenewables	http://www.scotrenewables.com	4.17
23	SeaPower	http://pa/cm200001/cmselect/ cmsctech/291/291ap25.htm	4.17
24	Sea Snail	http://www.rgu.ac.uk/cree	4.17
25	SMD Hydrovision	http://www.smdhydrovision.com	4.14
26	StarTide	www.starfishelectronics.co.uk	4.18
27	Statkraft	http://www.statkraft.com/pro/ energy_products	4.17
28	Swan Turbines	http://www.swanturbines.co.uk/	4.17
29	Swingcat	http://www.microhydropower.net/ mhp_group/portegijs	4.18
30	Tidal Generation	http://www.tidalgeneration.co.uk/index.html	4.18
31	Tidal Sails	http://www.tidalsails.com/	4.17
32	TidalStream	http://www.tidalstream.co.uk/	4.17
33	Tocardo Tidal Energy	www.tocardo.com	4.18
34	Underwater Electric Kite	http://uekus.com/	4.15
35	University of Southampton	http://pesn.com/2006/06	4.17
36	University of Strathclyde	http://www.mecheng.strath.ac.uk/ staffprofile.asp?id=37	4.18
37	University of Naples MYTHOS	www.dpa.unina.it/coiro/Marine-renewable-synthesis.pdf	4.17
38	Verdant Power	http://www.verdantpower.com/	4.16
39	Water Wall Turbines	http://www.wwturbine.com/	4.18
40	Wave Rotor	http://www.ecofys.nl/nl /expertisegebieden	4.18

references in the present chapter. In general, devices that, at the time of writing (2007), have been deployed at full scale or near full scale are detailed in Sections 4.2 to 4.16; other proposals are described in Section 4.17, and early ideas in Section 4.18.

4.2   Blue Energy

The Blue Energy ocean turbine consists of a three- or four-bladed vertical-axis rotor based upon the principles described variously as the Kobold (see also Ponte di'Archimede below) or Davis turbine (Chapter 3). According to Blue Energy (2008), the device was conceived by aerospace engineer Barry Davis, and represents two decades of Canadian research and development. The fixed hydrofoil blades are connected to a rotor that drives an integrated gearbox and electrical generator assembly. The turbine is mounted in a concrete marine caisson, which anchors the unit to the ocean floor, directs flow through the turbine, further concentrating the resource, and supports the gearbox and generator above it. The hydrofoil blades employ the hydrodynamic lift principle, which causes the turbine foils to move proportionately faster than the speed of the surrounding water, as detailed in Chapter 3. The blade angle is optimized to ensure that the rotation of the turbine is unidirectional on both the ebb and the flood of the tide.

The history of development of the device is detailed on the Blue Energy website. Preliminary tests were carried out on a vertical-axis water turbine in the National Research Council Hydraulics Laboratory test flume in Ottawa during 1981 and 1982. The preliminary experiments indicated that 'power outputs of the ducted turbine were up to 5 times greater than when the turbine was mounted in a freestream'. This rather astonishing statement is not explained further, nor supported by any publicly available data.

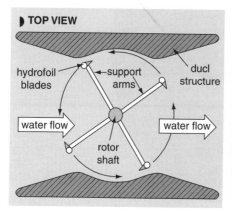

Figure 4.1   (Left) Plan view of Blue Energy's four-bladed vertical-axis rotor concept. (Right) The small-scale three-bladed test rig (from Blue Energy, 2008).

Blue Energy built a 20 kW vertical-axis turbine in early 1983 with a vertically mounted induction unit driven by a variable-speed gearbox. The device was installed in the Seaway of Cornwall in Ontario and connected to the Niagara Power Corporation grid. The device developed up to 20 kW on the grid. Blue Energy claim that the generator had an efficiency ranging from 26 to 52.5 %, and the efficiency of the whole system was said to be about 45 %.

4.3 Clean Current

Clean Current's tidal turbine generator (Figure 4.2) is a bidirectional, ducted, horizontal-axis turbine with a direct-drive, variable-speed, permanent magnet generator. Clean Current's website claims that 'this proprietary design delivers superior water to wire efficiency, a significant improvement over competing freestream tidal energy technologies'. Clean Current's turbine is a simple design in which only the rotor assembly and the blades move: the magnets for the generator are embedded in the blades, and the windings in the stationary casing. This is the same basic concept as is used by the OpenHydro turbine described below. Clean Current claim that 'the turbine generator has a design life of 10 years (major overhaul every 10 years) and a service life of 25–30 years'.

Clean Current's European Patent EP 1 430 220 B1 details a design concept whereby the 'rotor discs and generator are adapted to be removed as a modular unit such that maintenance and replacement is easily facilitated'.

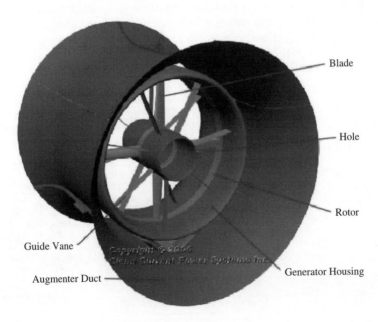

Figure 4.2 Clean Current's shrouded horizontal-axis turbine (from Clean Current, 2008).

Clean Current's direct-drive, variable-speed, permanent magnet generator incorporates features that allow the generator to be configured to produce either alternating or direct current. This permanent magnet generator is also specified in the European Patent.

Simple calculations have been used to test Clean Current's claim (Clean Current, 2008) that a 20 m blade diameter will deliver 2.2 MW at 3.0 m s^{-1}, and 3.5 MW at 3.5 m s^{-1}. Equation (2.5) shows that the overall power output, P_e, from a circular device is

$$P_e = \frac{\rho}{2} e \pi R^2 U^3 \tag{4.1}$$

so that, with a 40 % efficiency, the 20 m device will deliver 1.7 MW at 3.0 m s^{-1}, and 2.7 MW at 3.5 m s^{-1}. These figures are significantly less than Clean Current's claim. Alternatively, Clean Current's power outputs would require an efficiency of about 52 % to be achieved. If Clean Energy's figures are correct, the additional output is probably due to the effect of the ducting, which increases the capture area by some 50 %.

Clean Current's turbine has been successfully installed at the Race Rocks tidal energy project in conjunction with EnCana Corporation and a local educational institution, Pearson College.

4.4 Engineering Business

Engineering Business' Stingray (Figure 4.3) consists of a reciprocating hydroplane that has a variable angle of attack relative to the approaching water stream. This causes the supporting arm to oscillate, which in turn forces hydraulic cylinders to extend and retract, producing high-pressure oil that is used to drive a generator.

The company obtained funding from the British government's Department of Trade and Industry (DTI, now BERR) and from the Regional Development Agency's

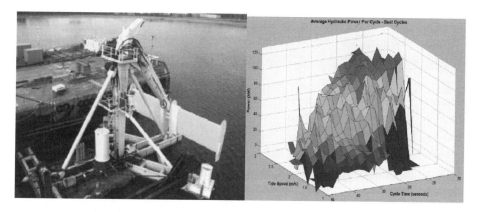

Figure 4.3 Engineering Business's Stingray ready for deployment in the Shetland Islands, and power output predictions (from Engineering Business, 2003).

New and Renewable Energy Centre (NaREC) to design and build the device which consisted of a single hydrofoil. The device was deployed for a period of months in Yell Sound, in the Shetland Islands of northern Scotland, in 2002. According to Engineering Business, the project delivered a mass of technical (Figure 4.3) and commercial data, and was completed on time and under budget. However, it resulted in the following statement:

> For a small private company to stay in business the golden rule is to generate cash. There is funding available for future development, but not on the scale or basis that will allow EB to rapidly or profitably make Stingray a commercial reality. Despite the significant support of the DTI, NaREC, and Shetland, given the timescales involved and investment required, EB cannot continue to sustain this project on a non-profit basis. Because of this we have taken the hard decision to put Stingray on hold.

Engineering Business (2003, 2008) suggest that the device oscillates with a period, T_o, of 24 s in a 2 m s^{-1} tidal stream and developed an electrical output of 117.5 kW. We use the potential output power equation

$$P_e = \frac{\rho}{2} eA \frac{U}{T_o} U^2 \tag{4.2}$$

Thus, an area, A, of 400 m^2 with $T_o = 24$ s suggests that the device's overall efficiency is only about 2 %! This is clearly due to the relatively slow oscillation speed, which effectively allows a great deal of the tidal stream to pass through the device without intersecting the hydrofoil. Pulse Generation (cf. Section 3.14) attempt to address this problem with a reciprocating design that oscillates much more quickly.

4.5 Gorlov helical turbine

The Gorlov helical turbine is a crossflow turbine with aerofoil-shaped blades, as shown in Figure 4.4 (upper and lower). The device provides a reaction thrust that can rotate the blades at twice the speed of the water flow (Bedard, 2005). The turbine is self-starting and, owing to its axial symmetry, rotates in the same direction during both the flood and the ebb of the tide. According to Bedard (2005), the standard unit has a diameter of 1.0 m, a length of 2.5 m, and a rating of 1.5 kW at 1.5 m s^{-1}, rising to 180 kW at 7.72 m s^{-1}, with an overall efficiency of about 30 %.

We can examine that claim with the crossflow potential hydraulic power equation (equation (2.5)) which states that

$$P_e = \frac{\rho}{2} eAU^3 \tag{4.3}$$

with $A = 2.5$ m^2 and $e = 0.3$, and $U = 1.5$ and 7.72 suggests outputs of 1.2 kW and 173 kW respectively, which are similar to the figures given above. However, it is not clear where a tidal current of 7.72 m s^{-1} can normally be found!

Figure 4.4 (Upper) The Gorlov helical turbine (Gorlov Helical Turbine, 2008). (Lower) Six-blade helical turbine installed at the research station of the Tide-Energy Project near the mouth of the Amazon (from Goreau *et al.*, 2005).

The turbine is said to have a basic cost of $6000. Goreau *et al.* (2005) state that the Gorlov helical turbine has been deployed at the 'research station of the Tidal-Energy Project near the mouth of the Amazon River'.

4.6 Hammerfest Strøm

Hammerfest Strøm (2008) utilizes its own water mill technology to convert the kinetic energy in a tidal current into electricity. The tidal stream turbines (Figure 4.5) are propellers with a 15–16 m blade diameter, mounted on a gravity base tower placed on the seabed. The current drives the propeller, with its blades automatically

Figure 4.5 Hammerfest Strøm AS's tidal turbine being deployed in the Norwegian fjords (from Hammerfest Strøm, 2008).

adjusted to their optimum orientation in the prevailing current. Each propeller is coupled to a generator from which the electricity is fed via a shore cable to a transformer and then into the grid. The prototype device was deployed on the seabed in Kvalsundet in Norway during early 2003.

Hammerfest Strøm's website makes the claim that the prototype device has an installed capacity of 300 kW and delivers 700 MW h per year to the grid, with pitch-controlled blades of 10 m. Thus, the mean power output is 700 000/8760 = 80 kW. Simple calculations have been used to test these figures using equation (2.5) for the overall power output, P_e, from a circular device:

$$P_e = \frac{\rho}{2} e \pi R^2 U^3 \tag{4.4}$$

For a 40 % efficiency and a mean current of 1.0 m s^{-1}, the 10 m radius device will deliver 88 kW. These figures generally support Hammerfest Strøm's claim.

4.7 Lunar Energy

The Lunar Energy Rotech tidal turbine (RTT) is based upon Rotech Engineering's concept, for which Lunar Energy has taken a worldwide, in-perpetuity licence. A team of experienced collaborators has been drawn together, including Rotech (lead designers and project managers), Atkins (structural design), ABB (generators), Hagglands and Bosch Rexroth (hydraulic pumps, motors and circuits), SKF (bearings), Garrad Hassan (control algorithms and hardware), and Wichita (hydraulic brake) to develop the concept.

Figure 4.6 Lunar Energy's RTT concept (from Lunar Energy, 2008).

The RTT is a bidirectional horizontal-axis turbine housed in a symmetrical Venturi duct. The Venturi operates as a diffuser and creates a low pressure behind the rotor, thus accelerating the flow into the duct. The symmetrical duct and rotor also allow the device to operate with equal efficiency in both the flood and the ebb currents. Power from the rotor is taken by high-pressure hydraulic fluid and accelerated onto an industry-standard turbine and generator. The resulting electrical output is then cabled ashore.

Lunar Energy (2008) states that the commercial RTT2000 is a 2 MW device that achieves its rated output in 3.0 m s^{-1} tidal flow through a rotor with a diameter of 32 m and will weigh 2500 t. At the time of writing, Lunar Energy had completed mathematical modelling and computational fluid dynamic (CFD) simulations and 1/20-scale tow tank testing. The CFD work and the tank testing is said to have demonstrated a turbine efficiency of 49 %, with an overall efficiency of about 40 %.

The initial capital costs of the RTT200 are estimated by Lunar Energy to be approximately £1.6–2.0 million, but are expected to reduce with learning by doing (LBD) and larger-scale cost reduction (LSCR) (Bedard, 2005). Maintenance costs will depend upon extracting a 'cassette' from the centre of the unit to the surface containing the rotor and all moving and electrical components.

4.8 Marine Current Turbines

The SeaGen technology under development by Marine Current Turbines (MCT) consists of twin axial-flow rotors of 15–20 m diameter, running at 10–15 rpm, each driving a generator via a gearbox much like a hydroelectric turbine or a wind turbine (Marine Current Turbines, 2008). The twin power units of each system are mounted on wing-like extensions either side of a tubular steel monopile some 3 m in diameter which is set into a hole drilled into the seabed. The technology for placing monopiles at sea is well developed by Seacore Ltd, a specialist offshore engineering company that is cooperating with MCT in this work. MCT's website claims that a unique, patented feature of the technology is that the turbines and accompanying power units can be raised bodily up the support pile clear above sea level to permit access for maintenance from small service vessels.

The first phase of the project was called SeaFlow and involved the design, construction, installation, and testing of the 300 kW single 11 m diameter rotor system off Lynmouth, Devon, UK. This was successfully installed in May 2003 and used a dump load in lieu of a grid connection. This phase cost £3.4 million and was financially supported by the partners together with the UK DTI, the European Commission, and the German government.

The second phase is called SeaGen and involves the design, manufacture, installation, and testing of the first full-size, grid-connected, twin-rotor system to be rated at 1.2 MW. This phase is expected to cost approximately £8.5 million, including

Figure 4.7 Artist's impression of Marine Current Turbines' SeaGen twin-impellor device (from Marine Current Turbines, 2008).

grid connection, and is financially supported by the operating partners and the UK DTI (UK BERR) who have awarded a grant of £4.27 million.

MCT's website claims that the prototype device has an installed capacity of 300 kW, with pitch-controlled blades of 11 m. Simple calculations have been used to test these figures using equation (2.5) for the overall power output, P_e, from a circular device:

$$P_e = \frac{\rho}{2} e \pi R^2 U^3 \qquad\qquad (4.5)$$

For a 20 % efficiency and a mean current of 2.0 m s^{-1}, the 11 m radius device will deliver 305 kW. These figures generally support MCT's data.

4.9   Neptune Renewable Energy

The Proteus Mark NP1000 vertical-axis tidal turbine was developed by Neptune Renewable Energy Ltd and tested by the University of Hull. The Mark I design consisted of a 5.00 m high, 5.00 m diameter, twelve-bladed, vertical-axis, crossflow rotor in the centre of a symmetrical, 22 m long × 14 m wide diffuser duct. The device was constructed at 1/10 scale and was tested in the tidal River Hull (Figure 4.8). CFD analysis was then used to optimize the rotor and duct design, as shown, for example, in Figure 4.9. Here, the torque on the rotor is plotted as the rotor turns through 360°. The first peak represents the flow as it strikes the blade, and then the second peak is due to the crossflow process as the fluid leaves the blade and exits the device. The red line shows the improvements introduced with the Mark II design.

In the Mark II version, the rotor was redesigned and the duct angles were reduced to 7° to prevent boundary layer separation. The single deflector plate was replaced by two sets of three vertical shutters to direct the flow at close to the optimum angle of 16° onto the rotor.

The patented Mark III device (Neptune Proteus, 2008) improved the rotor and duct design and utilized separate buoyancy chambers rather than the steel hulls of the earlier versions (Figure 4.10). The axle torque is taken by a right angle gearing to 2 × 200 kW permanent magnet DC generators and then via industrial regenerative drives and control systems to be cabled ashore and grid connected. Computational

Figure 4.8   (Left and middle) The Neptune Proteus Mark I at 1/10 scale. (Right) Testing in the tidal River Hull.

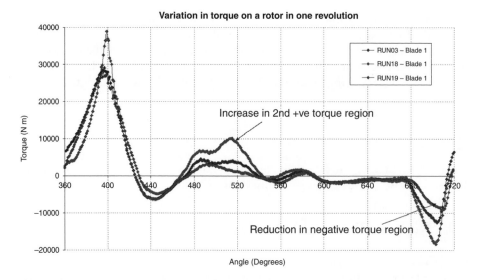

Figure 4.9 The CFD simulation of the shaft torque on the Neptune Proteus Mark I and Mark II devices.

fluid dynamics was used to optimize the rotor design and orientation and to promote the crossflow element which provides a positive torque both on entering and on exiting the rotor. The whole is moored fore and aft and operates equally efficiently on both the flood and the ebb tides. Control systems vary the shutter angles and the electrical load in order to maintain a tip speed ratio of about 0.35 for maximum efficiency.

Proteus is cost engineered for mesoscale, estuarine energy production (Figure 4.11) and is designed to generate electricity at both capital costs and operating costs that are significantly less than those of other tidal stream power devices and competitive with onshore wind. The basic functioning of the device is described by the transfer functions in Chapter 3:

$$P_H = \frac{\rho}{2} A_o U_0^3 \tag{3.3}$$

$$J_{VA} = 0.5 N' C_D e_{HY}^3 \frac{R_F}{R} \tag{3.37}$$

$$P_{SVA} = J \rho A T_{SR} U^3 (1 - T_{SR})^2 \tag{3.38}$$

Thus, equation (3.4) gives $P_H = 1350\,\text{kW}$ for an A_o of $15 \times 6.5 \approx 100\,\text{m}^2$ and $U = 3\,\text{m s}^{-1}$. Equation (3.37) gives $J \approx 10$ for $N' = 8$, $C_D = 2.3$, $e_{HY} \approx 1$, and $R_F/R = 1$. Equation (3.38) then gives $P_S = 452\,\text{kW}$ for $\rho = 1026$, $U = 3.0$, and $A = 12\,\text{m}^2$ and for the optimum $T_{SR} = 0.35$. Clearly, the device is theoretically and commercially viable and operates at an overall efficiency of 452/1300, or about 35 %. The device is expected to generate $1000\,\text{MW}$ h year^{-1} in a peak Spring tidal stream site of about $3.0\,\text{m s}^{-1}$. The first full-scale demonstrator is presently under construction.

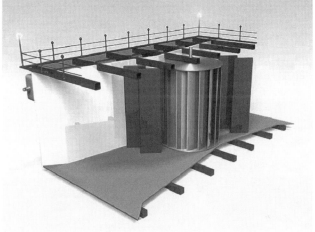

Figure 4.10 The Neptune Proteus Mark III, vertical-axis, crossflow rotor concept.

Figure 4.11 Low visual impact of a 5 MW (10 000 MW h year^{-1}) array.

4.10   OpenHydro

The open-centre turbine is an example of a simple idea proving itself to be one of the most effective solutions to the problem. Although the company plans to build larger devices, the prototype that was installed at the European Marine Energy Centre's (EMEC) site appears to have a diameter of about 6.0 m, so that the power rating (for a 40 % efficiency in a $3 \, \text{m s}^{-1}$ flow) is about (equation (4.1)) 50 kW.

The device (Figures 4.12 and 4.13) consists of a shaped duct, which improves the hydrodynamic efficiency of the rotor. The duct houses the stator which is fixed to the rotor and enclosed within the permanent magnet generator. The open centre increases the efficiency of the rotor as well as providing 'an escape route for marine life'.

Although open-centre technology is unique and covered by a suite of world-wide patents, the concept is similar to the devices described by Clean Current (cf. Section 4.3) and the University of Southampton (cf. Section 4.17.12).

The OpenHydro website (OpenHydro, 2008) explains the origins of the company and technology as follows: 'OpenHydro was formed in 2004 following the negotiation for the purchase of the world technology rights to the Open-Centre Technology (by an Irish consortium). The technology had been developed by Irish American, Herbert Williams, in the U.S. over the previous 8 years'.

Figure 4.12   OpenHydro's tidal turbine installed at the European Marine Energy Centre in the Orkney Islands, Scotland (from OpenHydro, 2008).

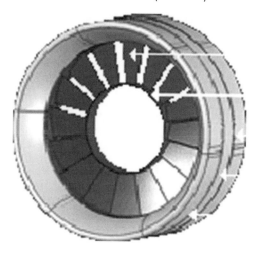

Figure 4.13 OpenHydro's tidal turbine (from http://www.openhydro.com/news/
OpenHydroPR-140107.pdf)

4.11 Ponte di Archimede (Kobold)

Calcagano *et al.* (2007) report that the Kobold turbine utilized in this device is
a straight-blade, vertical-axis hydroturbine (Ponte di Archimede, 2008). The prototype had three blades, each 5 m long with a chord of 0.4 m on a 6 m diameter
structure, yielding Reynolds numbers from 800×10^3 to 2×10^6 (Figure 4.14).
The blade angles are controlled by a series of levers in order to maintain an optimum angle of attack. The rotor is mounted on a 10 m diameter platform and
deployed 150 m offshore in the Straits of Messina between Italy and Sicily in the
Mediterranean. The device is moored with four blocks each of 350 kN.

Figure 4.14 (Left) The three-bladed Kobold turbine. (Right) The test rig generating
power in the Straits of Messina (from Ponte de Archimede, 2008).

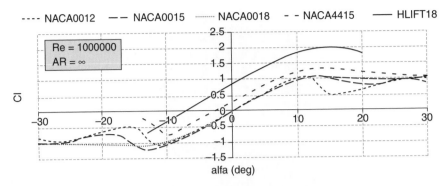

Figure 4.15 Comparison between the lift from the H-lift and classical airfoils (from Ponte de Archimede, 2008).

The rotor is connected through a 90:1 gearbox to a brushless, three-phase, four-pole generator. The development team measured the overall device efficiency at 23 % in accordance with the definition given by equation (3.4):

$$e = \frac{P_E}{P_H} \qquad (3.4)$$

Figure 4.15 shows their results for the lift coefficient generated by the different aerofoil sections in laboratory tests at various angles of attack. As discussed in Chapter 3, the lift increases to a peak value of about 2 for an angle of attack of about 15°.

4.12 Pulse Generation

Pulse Generation's device uses similar principles to Engineering Business's Stingray, but uses a patented mechanism to cause pairs of horizontal hydrofoils (shown in Figure 4.16) to undergo vertical reciprocating movements at a much higher frequency than Stingray. The hydrofoils are connected through solid levers or high-pressure hydraulic fluid linkages to a gearbox and generator housed in a steel nacelle above the waterline and easily available for servicing and maintenance.

The device has been conceived for shallow water deployment, where wave activity is limited, leading to capital and operational cost reductions. In addition, Pulse Generation's website (Pulse Generation, 2008) claims that changing the amplitude of oscillation of the foils means the system can be adjusted for different flow depths so that the deeper water available at high tide can be exploited.

The system, in common with the vertical-axis rotors such as Neptune Proteus or the Kobold turbine, takes energy from a rectangular cross-section of water which provides more hydraulic power per unit width of the bed than is available to circular turbines with the same diameter as the length of the hydrofoils. For example, the pulse stream device, or the vertical-axis crossflow rotor with sides of $2R$, has a total area of $4R^2$. A comparable circular turbine has a cross-sectional area of πR^2. Thus,

Humber Prototype Design

Figure 4.16 Pulse Generation's pulse stream 100 kW device (from Pulse Generation, 2008).

the pulse stream or crossflow has some $4/\pi = 1.28$ or nearly 30 % greater potential hydraulic power per unit bed width.

Shallow flows are an ideal resource because they are close to the shore and are often sheltered from heavy seas. This leads to major cost reductions. The pulse system can couple any number of foil groups to a single electricity generator. This means that machines can be developed for any depth of water by adding the appropriate number of foils.

Pulse Generation has been funded by a £900 000 grant from the British government's Department of Trade and Industry (DTI, recently rebranded BERR) towards the £2 million total cost to build a scale 100 kW prototype for deployment in the Humber Estuary on the coast of eastern England. A 1/10 scale model was tested in the University of Hull recirculating flume as one element of this project.

4.13 SeaPower

Petterson and Langström (2003) report that the patented EXIM tidal and stream turbine was originally designed by Langström, under the umbrella of EXIM Strömturbiner (then taken over by Swedish-owned SeaPower Scotland), with the objective of developing a method for converting the kinetic energy of ocean currents into rotary energy. A number of turbine models were originally trialled, before the Savonius rotor was adopted for further development. The concept now consists of a stacked set of turbines, each measuring about 1 m high with a diameter of about 1.5 m, driving a generator. Full-scale tests were performed at the Ship Design and Research Centre in Gdansk, Poland, in August 2002, in cooperation with a partner, Navimor Sp. z o.o., as a joint venture known as Seapower Polska Sp.

Further full-scale tests were carried out in 2003 in Bluemell Sound off the Shetland Islands, Scotland, from Delta Marine's *Voe Venture* (Figure 4.17). Teknik & Service of Sweden supplied the generator and gearbox, and the turbine with supporting frames was manufactured by Ocean Kinetics of Lerwick. During testing, the prototype was attached to the anchored ship and the electrical output was measured.

When the turbine was tested at a tidal-current speed of around $2\,\mathrm{m\,s^{-1}}$, the induction generator delivered 2 kW at 48 rpm, rising to around 3 kW at $2.5\,\mathrm{m\,s^{-1}}$.

Petterson and Langström (2003) then calculate that the maximum power output from two turbines of 2 m diameter and 9 m length in a tidal current of $3.3\,\mathrm{m\,s^{-1}}$ would be 40–50 kW.

We can examine that claim with the potential hydraulic power equation, which states that:

$$P_{\mathrm{e}} = \frac{\rho}{2} eHWU^3 \tag{4.6}$$

The width $W = 2.0\,\mathrm{m}$, height $H = 9.0\,\mathrm{m}$, flow $U = 3.3\,\mathrm{m\,s^{-1}}$, and an output of 50 kW suggests an efficiency of only about 15 %.

Figure 4.17 SeaPower's vertical-axis rotor device (from Petterson & Langström, 2003).

4.14 SMD Hydrovision

REUK (2007) reports that SMD Hydrovision's TidEl differs from most tidal turbines in that it floats and is restrained by mooring chains to the seabed rather than being fixed to piles driven into the seabed. The full-size prototype will incorporate two buoyant 500 kW generators joined together by a crossbeam, giving a total power capacity of 1 MW (at 2.3 m s^{-1}). The fixed-pitch turbine blades will be 15 m in diameter and consist of three 8 m blades on a 2.5 m hub. The blades are variable pitch and controlled to maintain the optimum angle of attack at various flow speeds. The anchoring mechanism leaves the device free to align itself downstream of the tidal flow as the tide changes direction. It is claimed that TidEl can be installed in deep waters (below the light depth), and so barnacle and other sealife damage can be avoided, reducing maintenance intervals and downtime.

A 1/10-scale prototype system, including the generator, mooring, and grid integration controls, was tested at NaREC with part funding from a DTI grant (Figure 4.18). Full-scale testing in the Orkneys is planned and part funded by a further DTI grant of £2.7 million (out of a project cost of £4.5 million).

Calcagano *et al.* (2007) report that the blade and drive machinery efficiencies of this device at 1.0 and 2.0 m s^{-1} are 35 and 55 % and 35 and 81 %, giving overall efficiencies of 17 % and 27 % respectively.

Hydrovision estimates that the total capital cost of the 100 MW farm would be around £140 million, with £5 million required per year for management and maintenance, and a further £14 million required for decommissioning.

Figure 4.18 SMD's TidEl device in close up (from SMD Hydrovision, 2008).

Simple calculations have been used to test SMD's claim (SMD Hydrovision, 2008) that a 15 m blade diameter will deliver 1 MW at 2.3 m s^{-1}. Equation (2.5) shows that the overall power output, P_e, from a circular device is

$$P_e = \frac{\rho}{2} e \pi R^2 U^3 \tag{4.7}$$

so that, with a 25 % efficiency, the device will deliver 1.08 MW, which is in accordance with SMD's claim.

4.15   Underwater electric kite

Bedard (2005) reports that the underwater electric kite (UEK) (Underwater Electric Kite, 2008) is a twin, high-solidity (85–95 %), horizontal-axis device with 'augmentor rings' (ducts), which increases the internal velocity of the flow

Figure 4.19 UEK System's device (from Underwater Electric Kite, 2008).

Table 4.2 Potential hydraulic power, electrical power, and efficiency for the underwater electric kite

	P_H (kW)	P_e (kW)	e
1.543	114	74	0.65
2.058	271	149	0.55
2.469	468	261	0.56
2.984	826	447	0.54

(Figure 4.19). A 3.3 m unit was tested at Ontario Hydro's DeQew Hydroelectric Power Plant, and a 5.5 m unit was tested in Chesapeake Bay. The augmentor is said to influence flow over a diameter of 6.2 m, and a rotor diameter of 4.0 m has a rated power of 400 kW at 3 m s^{-1}.

The detail of the augmentor ring is that it flares out behind the turbine, creating a low-pressure zone that enhances flow through the device. Thus, the augmentor is operating as a diffuser (Chapter 3). The power generated from the shrouded turbine drives a brushless alternator through a planetary gearbox. The variable alternator output is connected to a rectifier and filters, and is controlled by the transmission regulator.

The mooring is said to be simply fore and aft anchors, and the device is symmetrical, working equally well in either flood or ebb currents. Power is delivered to the shore as DC and inverted to a grid synchronous supply.

Experiments with a single device on a moving boat generated the data shown in Table 4.2 on an area of 38.6 m^2 for the augmentor ring.

The resulting overall efficiency is about 55–65 %.

4.16 Verdant Power

Bedard (2005) reports that the Verdant Power kinetic hydropower (VPKHP) system is a three-bladed, axial-flow turbine that incorporates a patented blade design having a high efficiency over a range of flow speeds. The turbine rotor drives a speed increaser, which drives a grid-connected, three-phase induction generator. The gearbox and generator are in a waterproof, streamlined nacelle which is mounted on a streamlined pylon (Figure 4.20). The pylon assembly has internal yaw bearings allowing it to pivot the turbine with the ebb and flood of the tidal current. The pylon is bolted to a pile fixed to the seabed. Underwater cables carry the AC current ashore for connection to the grid using standard distributed generation switchgear.

Verdant turbines are being installed in the East River, New York, where the 5 m diameter rotor is rated at 39.5 kW in a tidal current of 2.2 m s^{-1}. The data shown in Table 4.3 are provided by Calcagano *et al.* (2007) for a cross-sectional area of 19.6 m^2.

These data suggest an overall efficiency of 35–40 %.

Figure 4.20 Verdant Power's kinetic hydropower system (from Verdant Power, 2008).

Table 4.3 Potential hydraulic power, electrical power, and efficiency for the Verdant Power turbine

	P_H (kW)	P_e (kW)	e
1.00	10	4	0.40
1.50	34	12	0.35
2.00	80	29	0.36

Simple calculations have been used to test Verdant's claim that a 5 m blade diameter will deliver 39.5 kW at 2.2 m s^{-1}. Equation (2.5) shows that the overall power output P_e from a circular device is

$$P_e = \frac{\rho}{2} e \pi R^2 U^3 \tag{4.8}$$

so that, with an efficiency of 35–40 %, the device will deliver 36–42 kW, which is in accordance with Verdant's claim.

Verdant (Verdant Power, 2008) estimates a production cost of \$100 000 for a one-off turbine, decreasing by 20 % for volume manufacture. This is equivalent to

$100\,000 \times 40/1000 = \$2.5$ million $\mathrm{MW^{-1}}$, which is in excess of some of the other systems described here.

4.17 Other proposals

4.17.1 BioPower Systems

Few details and no published data are available on this device, but the tidal energy conversion system, bio*STREAM*™ (Figure 4.21 (left)) is apparently 'based on the highly efficient propulsion of Thunniform-mode swimming species, such as shark, tuna, and mackerel' (BioPower Systems, 2008). The motions, mechanisms, and caudal fin hydrofoil shapes of such species are said 'to be up to 90 % efficient at converting body energy into propulsive force'. The bio*STREAM*™ mimics the shape and motion characteristics of these species but is a fixed device in a moving stream. In this configuration, the propulsion mechanism is reversed and the energy in the passing flow is used to drive the device motion against the resisting torque of the O-*DRIVE*™ electrical generator. Owing to the single point of rotation, this device can align with the flow in any direction, and can assume a streamlined configuration to avoid excess loading in extreme conditions. Finally, bio*STREAM*™'s website claims that systems are being developed for 500 kW, 1000 kW, and 2000 kW capacities to match conditions in various locations. This may prove difficult because, if we use the potential hydraulic power equation, then the device would have to be a massive 65 m high with a 20 m sweep to achieve the last named output in a stream of $2\,\mathrm{m\,s^{-1}}$ at 40 % efficiency. There are few if any sites of such depth with such currents. Another problem stems from the physics of the device. Shapes (such as fishtails) that may be very efficient at converting oscillating motions into propulsive forces because of the added mass generated in the wake and against which the fin reacts are necessarily inefficient when operating in the opposite sense because the added mass is simply a drain on the system.

Figure 4.21   (Left) BioPower's BioStream concept (from BioPower Systems, 2008). (Right) Bourne Energy's TideStar concept (from Bourne Energy, 2008).

4.17.2 Bourne Energy

According to their website (Bourne Energy, 2008), each of Bourne Energy's River-Stars is a 20 ft (6.7 m) long, self-contained energy module composed of a stabilizer, an energy absorber (a double four-bladed turbine), an energy transmission, a mooring system, and an energy conversion and control system designed to be sited in long arrays across the tide. Each unit, utilizing a proprietary turbine design, produces approximately 50 kW at peak capacity (Figure 4.21 (right)).

No further details are available, and it is not clear what type of blade configuration or generator is envisaged. The device appears to be moored, and, again, it is unclear as to whether the device reverses or is of symmetrical design to accommodate tidal flood and ebb currents. No performance data or dimensions are presently available.

4.17.3 Edinburgh University (POLO)

The Edinburgh University POLO was conceived by one of the early marine energy luminaries, Professor Stephen Salter, who, in the 1970s and 1980s, led the British development of wave power systems, principally Salter's oscillating DUCK. Salter (1998) and Edinburgh POLO (2008) describe the POLO as consisting of three sets of symmetrical vertical blades, each typically 6.7 m long with 1.9 m chords, supported at both ends by bearings in horizontal rings (Figure 4.22 (left)). This gives a large reduction in bending moments and bearing loads relative to a cantilevered support and allows a much greater total depth.

The blades have variable pitch by rotating about spars which join the horizontal rings and achieve a performance coefficient of over 0.4, which is comparable with vertical-axis wind machines. The rotor rings are braced between the spars by diagonal ties of high-tensile steel cable fitted with rotating fairings to minimize drag.

Figure 4.22 (Left) Edinburgh University's tidal rotor concept (from Edinburgh POLO, 2008). (Right) Hydrohelix Energy's concept drawing (from HydroHelix Energy, 2007).

The upper part of the rotor drives a ring-cam of the same diameter running through a floating torus which will be at least 3.4 m in diameter if maintenance staff are to stand comfortably.

4.17.4 Hydrohelix Energy

According to their website, Hydrohelix Energy has been developing a 'hydrolienne' technology for 7 years that generates electricity from marine tidal currents (Hydrohelix, 2007).

The work appears to be related to a political initiative to establish the Brittany region as a centre for marine renewables. Working within a consortium, the MARENERGIE project has been established for the realization of this goal. Images of the device suggest a line of seabed-mounted, ducted turbines (Figure 4.22 (right)) with an initial target installation of 1 MW. The device will be installed off the coastline of Brittany. At the same time, a very small-scale demonstrator named SABELLA with a capacity of 120 kW is also planned (Hydrohelix, 2007).

The work appears to coincide with an announcement by the French government (J.O. du 22 avril 2007) that there will be a purchase tariff on marine renewable energy of 15 c€ kW h^{-1} (equivalent to 150 € MW h^{-1}), which will 'make the exploitation of marine renewable resources attractive'.

4.17.5 Robert Gordon University

The Sea Snail structure that is presently under development at Robert Gordon University, Scotland, utilizes a number of upturned hydrofoils mounted on a frame to induce a downforce from the stream flow and anchor the device (Robert Gordon University, 2007) (Figure 4.23 (left)). The underlying theory was clearly demonstrated, and accurate records of applied forces taken, by means of a small-scale device (the Winkle), tested in a local watercourse.

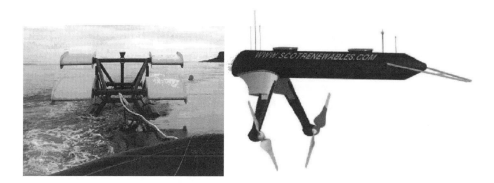

Figure 4.23 (Left) Robert Gordon University's Sea Snail (from Robert Gordon University, 2007). (Right) ScotRenewables' tidal turbine concept (from ScotRenewables, 2008).

The 'Winkle' was equipped with load cells and flow measurement transducers, the data from which gave a close correlation with the Sea Snail calculations. The Sea Snail was designed using a readily available parametric modelling package. This permits all dimensions, fits, and tolerances to be checked prior to full engineering drawing production. The concept was designed for ease of fabrication using standard steel sections and was made in kit form for transportation. The Sea Snail can be assembled in a few days by a small crew and a 60 t crane.

4.17.6 ScotRenewables

According to their website, the ScotRenewables tidal turbine (SRTT) system is a free-floating rotor-based tidal-current energy converter that began development from a PhD project in 2004 (ScotRenewables, 2008). The design has won various awards, including an Enterprise Fellowship – awarded by the Royal Society of Edinburgh and Scottish Enterprise.

The concept in its present configuration involves dual counter-rotating horizontal-axis rotors driving generators within subsurface nacelles, each suspended from separate keel and rotor arm sections attached to a single surface-piercing cylindrical buoyancy tube (Figure 4.23 (right)). The device is then anchored to the seabed via a yoke arrangement and compliant mooring system. A separate flexible power and control umbilical then connects to a subsea junction box. The rotor arm sections are hinged to allow each two-bladed rotor to be retracted so as to be parallel with the longitudinal axis of the buoyancy tube, giving the system a transport draught of under 4.5 m at full scale to facilitate towing the device into harbours for major maintenance.

No power data are presently available for the ScotRenewables tidal turbine.

4.17.7 Statkraft

According to their website (Statkraft, 2008), Hydra Tidal signed an agreement with Statkraft for the joint development and construction of a full-scale demonstration plant known as the MORILD project in 2004. The main project objective was to construct, test, and operate a working demonstration plant at a representative site in Kvalsundet in Tromsø, Norway. Power output to the grid should be about 1 MW. The proposed device appears to consist of a pair of twinned rotors mounted beneath a surface float and moored to the seabed (Figure 4.24 (left)),

4.17.8 Swan Renewables

The Swan Renewables tidal turbine (Swan Turbines, 2008) (Figure 4.24 (right)) was developed in Swansea University in Wales, and, according to their website, the device consists of a traditional turbine on an extendable underwater mounting that orients the device into the prevailing flow on a gravity base. The gearless, low-speed generator offers 'a high efficiency over a range of speeds with minimal maintenance demands through the use of novel structural and electromagnetic topologies'. A simple yawing mechanism is said to be used for maximum flow capture.

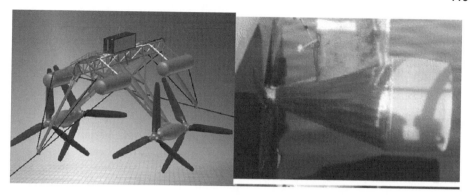

Figure 4.24 (Left) The Stakraft tidal turbine concept (from Stakraft, 2008). (Right) Swan Turbines' concept (from Swan Turbines, 2008).

No design dimensions or performance data appear to be available at the time of publication.

4.17.9 Tidal Sails

According to their website (Tidal Sails, 2008), Tidal Sails AS plans to develop, patent, and put on the market its power plant for the production of electricity from tidal currents. The principle is said to be comparable with a set of vertical blinds under water that are being pulled from start to end station by the tidal current (Figure 4.25 (left)). A magazine at the start station deploys sails at certain intervals, at the same time as the end station magazine is detaching and collecting arriving sails. The sails are fixed to long cables under water, and are pulled by the tidal stream. The cables are attached to a generator, which in turn produces electricity.

No dimensional or performance data were available for this concept at the time of publication. However, the device faces some fairly formidable design problems. The sails would be most efficient if they moved in a downtide direction, but then each sail would shield all of those behind it from the force of the flow for a very

Figure 4.25 (Left) Tidal Sail's concept (from Tidal Sails, 2008). (Right) TidalStream's concept (from TidalStream, 2008).

considerable distance downstream, so that the overall efficiency would be very low. If the sails formed an angle with the flow, so that the sails 'broad reached', then there would be engineering problems dealing with flow reversal.

4.17.10   TidalStream

TidalStream's website (TidalStream, 2008) explains that its concept has two 20 m rotors, is rated at 1–2 MW depending on current speed, and operates in 30–50 m water depths (Figure 4.25 (right)). Each rotor runs in clean water upstream of its support arm with a gravity base and the swinging arm ball joint. The swinging arm is hinged at its upper end to the main spar buoy so that it can be stowed for installation or removal. During operation it is held in place by a cranked strut. No performance data were available at the time of publication.

4.17.11   University of Naples MYTHOS

The MYTHOS turbine (Figure 4.26 (left)) is a development of the Kobold approach described earlier (see the Pointe d'Archimede section) and is said to consist of a vertical-axis crossflow turbine with straight blades (University of Naples, 2008). It apparently shows an improved overall efficiency due to the hydrodynamics of the blades and the direct-drive, permanent magnet generator.

4.17.12   University of Southampton

The University of Southampton design concept appears to be similar to that of OpenHydro, in that there is no separate gearbox and generator; instead, the electrical machinery is embedded in the surrounding duct (University of Southampton, 2008) (Figure 4.26 (right)). According to their website, the design does not need to turn around to be oriented into the flood and ebb currents because the shape of its

Figure 4.26   (Left) University of Naples' MYTHOS concept (from University of Naples, 2008). (Right) University of Southampton's 0.25 m diameter model (from University of Southampton, 2008).

turbine blades means that they turn equally well, regardless of which way the water flows past them. The blades are said to be placed in a specially shaped housing that channels the water smoothly through the turbine.

4.18 Early-stage proposals

In addition there are 13 very early-stage technologies:

(a) **Aquantis.** Bedard (2005) mentions this device by a subsidiary of Clipper Technologies who have announced that they are 'developing an in-stream tidal kinetic energy device, but are unable to share information about the device as a patent is pending'.

(b) **Atlantisstrom** (2008) describes a transverse horizontal-axis device with five drop-shaped blades, each 20 m long, between two 10 m diameter endplates. At the bottom of their rotation the blades are normal to the flow and generate torque, while at the top they move parallel with and against the flow. Estimated output is about 80 kW in a current of $3\,\mathrm{m\,s^{-1}}$. A similar device is described in outline by Isis-Innovation (2008).

(c) **C-Power.** An English company developing as yet unspecified tidal generation projects.

(d) **EvoPod.** This device is a floating horizontal-axis tidal turbine (EvoPod, 2008) developed by Overberg in north-east England.

(e) **Hydroventuri.** Although often mentioned, this is, strictly speaking, a tidal head power device consisting of a tube and pressure take-offs (Hydroventuri, 2008).

(f) **OcGen.** The Ocean Renewable Power Company (ORPC) has proposed the OcGen twin, horizontal-axis, crossflow turbine (ORPC OcGen, 2008).

(g) **StarTide** have proposed a very large-scale set of turbines on a single driveshaft linked to an Archimedean screw-type compressor that drives pressurized water into a turbine linked to a generator. The device appears to have very low driveshaft efficiency (StarTide, 2008).

(h) **Swingcat.** Jan Portegijs has proposed a catamaran-based, twin, horizontal-axis tidal turbine called the Swingcat (Swingcat, 2008).

(i) **Tidal Generation Limited** (TGL) are developing a 1 MW, fully submerged, horizontal-axis tidal turbine and claim innovative concepts for the installation and maintenance of the machine in deep water. The turbine is also claimed to be cheap to construct and easy to install owing to the lightweight ($80\,\mathrm{t\,MW^{-1}}$) support structure (Tidal Generation, 2008).

(j) **Tocardo Tidal Energy** have proposed a two-bladed horizontal-axis rotor and completed 'full-scale' tests with the 2.8 m diameter 35 kW device in 2005 (Tocardo Tidal Energy, 2008). The company appears to be planning a 10 m device to be known as the Tocardo Offshore.

(k) **University of Strathclyde** in Scotland have reported the development of a contrarotating marine current turbine.

(l) **Water Wall Turbines** (2008) have suggested that more than 40 MW is achievable with each of their units from tidal currents. No further details are given.

(m) **Wave rotor** is a very early-stage design concept consisting of three vertical-axis lift blades beneath a moored buoy. The name may be confusing as the lift will be provided by tidally induced, not wave-induced, currents. Details are at Wave Rotor (2008).

4.19 Bibliography

Atlantisstrom (2008) http://www.atlantisstrom.de/index_english.html [accessed 17 April 2008].

Bedard, R. (2005) *Survey and Characterisation of Tidal In Stream Energy Conversion (TISEC) Devices.* Report EPRI TP 004 NA. 185 pp.

BioPower Systems (2008) http://www.biopowersystems.com [accessed 18 March 2008].

Blue Energy (2008) http://bluenergy.com/ [accessed 18 March 2008].

Bourne Energy (2008) http://bourneenergy.com/ [accessed 18 March 2008].

Calcagano, G., Salvatore, F., Greco, L., *et al.* (2007) *Experimental and Numerical Investigation of a Very Promising Technology for Marine Current Exploitation: the Kobold Turbine.* http://www.pontediarchimede.it/shared_uploads/Download/13_DOCUMENTO.pdf [accessed 26 March 2008].

Clean Current (2008) http://www.cleancurrent.com [accessed 18 March 2008].

Edinburgh POLO (2008) http://www.mech.ed.ac.uk/research/wavepower/ [accessed 18 March 2009].

Engineering Business (2003) *Project Summary. Stingray Tidal Stream Energy Device. Phase 3.* http://www.engb.com/downloads/M04-102-01.pdf accessed 070520.

Engineering Business (2008) http://www.engb.com/services_09a.php [accessed 18 March 2008].

EvoPod (2008) http://www.oceanflowenergy.com/ [accessed 18 March 2008].

Goreau, T.J., Anderson, S., Gorlov, A., & Kurth, E. (2005) *Tidal Energy and Low-head River Power: a Strategy to Use New, Proven Technology to Capture these Vast, Non-polluting Resources for Sustainable Development. Submission to the UN.* http://www.globalcoral.org/Tidal%20Energy%20and%20Low-Head%20River%20Power.htm [accessed 16 April 2007].

Gorlov Helical Turbine (2008) http://www.gcktechnology.com [accessed 18 March 2008].

Hammerfest Strøm (2008) http://www.e-tidevannsenergi.com/ [accessed 18 March 2008].

Hydrohelix Energy (2007) http://cci-entreprises.icomme.fr/public/stand.php?id_stand=418 [accessed 4 June 2007].

Hydroventuri (2008) http://www.hydroventuri.com [accessed 18 March 2008].

Isis-Innovation (2008) http://www.isis-innovation.com/licensing/3325.html [accessed 6 September 2008].

Lunar Energy (2008) http://www.lunarenergy.co.uk/ [accessed 18 March 2008].

Marine Current Turbines (2008) http://www.marineturbines.com [accessed 18 March 2008].

Moore, J. (2007) *Quantification of Exploitable Tidal Resource in UK Waters.* ABPMer Report for the Juice Fund.

Neptune Proteus (2008) http://www.neptunerenewableenergy.com [accessed 18 March 2008].

OpenHydro (2008) http://www.openhydro.com [accessed 18 March 2008].

ORPC OcGen (2008) http://www.oceanrenewablepower.com/ [accessed 18 March 2008].

Pettersson, I. and Langström, G. (2003) *Northern Exposure: Testing Tidal Turbines*. International Water Power.

Pont di Archimede (2007) http://www.pontediarchimede.it/language_us/progetti_det.mvd?RECID=2& CAT=002&SUBCAT=&MODULO=Progetti_ENG&returnpages=&page_pd=p [accessed 18 April 2007].

Ponte di Archimede (2008) Ponte di Archimede (2007) [accessed 18 March 2008].

Pulse Generation (2008) http://www.pulsegeneration.co.uk/ [accessed 18 March 2008].

Pure Energy Systems (2007) http://peswiki.com/index.php/Directory:Tidal_Power#Technologies [accessed 16 April 2007].

REUK (2007) http://www.reuk.co.uk/TidEl-Tidal-Turbines.htm [accessed 21 May 2007].

Robert Gordon University (2007) http://www.rgu.ac.uk/cree/general/page.cfm?pge=10769 [accessed 18 April 2007].

Salter, S.H. (1998) Proposal for a large, vertical-axis tidal stream generator with ring-cam hydraulics, in *Proceedings of Third European Wave Energy Conference*, Patras, Greece, 30 September–2 October 1998.

ScotRenewables (2008) http://www.scotrenewables.com [accessed 18 March 2008].

SeaPower (2008) http://pa/cm200001/cmselect/cmsctech/291/291ap25.htm [accessed 18 March 2008].

SMD Hydrovision (2008) http://www.smdhydrovision.com [accessed 18 March 2008].

StarTide (2008) www.starfishelectronics.co.uk [accessed 18 March 2008].

Statkraft (2008) http://www.statkraft.com/pro/energy_products [accessed 18 March 2008].

Swan Turbines (2008) http://www.swanturbines.co.uk/ [accessed 18 March 2008].

Swingcat (2008) http://www.microhydropower.net/mhp_group/portegijs [accessed 18 March 2008].

Tidal Generation (2008) http://www.tidalgeneration.co.uk/index.html [accessed 18 March 2008].

Tidal Sails (2008) http://www.tidalsails.com/ [accessed 18 March 2008].

TidalStream (2008) http://www.tidalstream.co.uk/ [accessed 18 March 2008].

Tocardo Tidal Energy (2008) www.tocardo.com [accessed 18 March 2008].

Underwater Electric Kite (2008) http://uekus.com/ [accessed 18 March 2008].

University of Naples (2008) www.dpa.unina.it/coiro/Marine-renewable-synthesis.pdf [accessed 18 March 2008].

University of Southampton (2007) http://pesn.com/2006/06/13/9500281_Southampton_Tidal_Gene rator/[accessed 16April 2007].

Verdant Power (2008) http://www.verdantpower.com/ [accessed 18 March 2008].

Wave Rotor (2008) http://www.ecofys.nl/nl/expertisegebieden [accessed 18 March 2008].

Water Wall Turbines (2008) http://www.wwturbine.com/ [accessed 18 March 2008].

5

Modelling tidal stream power

5.1 Introduction

This chapter introduces the harmonic method for the synthesis of tidal stream power time series (cf. Chapters 1 and 3). The harmonic method was developed in the nineteenth century and has been used for tidal streams by, for example, Moore (2007a,b) and Hardisty (2007). In the harmonic method, the constructive and destructive interference of a number of components of the tide are synthesized in order to generate a year-long time series, at hourly intervals, of the tidal currents.

It is noted that a number of other methodologies have been reported with application to specific tidal areas. For example, ESRU (2007) describe an analysis of tidal streams around Scotland, and EPRI (2005) present an analysis of tidal streams around North America.

The objective of this chapter is to develop and explain the harmonic analysis of the tidal resource methodology for estimating the tidal power that may be extracted

The Analysis of Tidal Stream Power   Jack Hardisty
© 2009 John Wiley & Sons, Ltd

from a new site, based upon limited or minimal flow information. The methodology, which we will call the *simplified tidal economic model* (STEM), has five stages:

1. **Tidal data sites**. Identify representative sites for flow analysis.

2. **Tidal diamonds**. Acquire maximum Spring and Neap, flood, and ebb surface current speeds and water depths at the chosen site. In certain cases (sites with a mixed, diurnal, and semi-diurnal tide), estimate the Formzahl number (cf. Section 5.5.2).

3. **Harmonic decomposition**. Determine current amplitudes for the six harmonics:

   (a) M_{U2} lunar semi-diurnal;
   (b) S_{U2} solar semi-diurnal;
   (c) M_{U4} lunar quarter-diurnal (also known as the shallow water effect);
   (d) K_{U2} lunisolar semi-diurnal;
   (e) K_{U1} lunisolar diurnal;
   (f) O_{U1} principal lunar diurnal.

4. **Simplified tidal economic model**. Use the STEM to synthesize a 12 month time series of hourly current speeds.

5. **Potential hydraulic power**. Use the synthetic time series to determine representative statistics and the potential hydraulic power at the site.

This chapter is based around the author's STEM software, which is available from the book's website. STEM performs the calculations involved in steps 3 to 5 and provides an estimate of the electrical output and power cost from information about the device as described below.

5.2   Global tidal streams

It is not easy to take a global perspective on tidal-current speed distribution because the flows are strongly modulated by the local topography. Thus, the constrictions of a tidal channel or of a headland will usually be responsible for the strongest tidal currents, while a short distance away, outside the influences of the constriction, much lower flow speeds are experienced. There are, however, some basic principles that control the global distribution of the areas of highest current speed. On a global scale, as explained in earlier chapters, the world ocean oscillates with tidal waves in response to the gravitational forcing, as shown in Figure 5.1. Large tidal ranges occur on the north-west coasts of North and South America, Europe, Australia, and New Zealand and on the east-central coasts of Africa, Asia, and South America, along with other, more localized, areas.

The result is a series of nodal and antinodal points, so that maximum flows occur in the areas that are distal from the amphidromes and, generally, have the largest tidal ranges. The general distribution of these currents is shown in Figure 5.2.

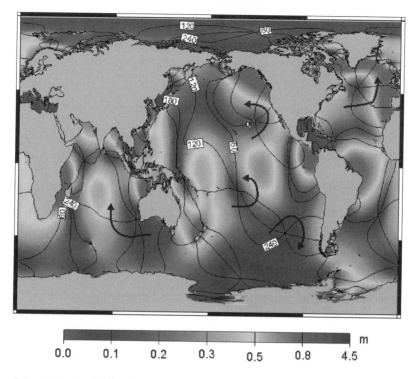

Figure 5.1 Global distribution of tidal range and amphidromic points (from Chalmers University, 2008). A URL link to the full colour original of this image is available at the book's website.

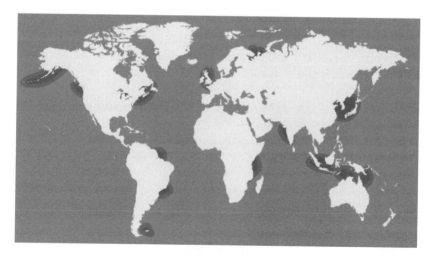

Figure 5.2 Global distribution of strong tidal streams (from OpenHydro, 2008). A URL link to the full colour original of this image is available at the books website.

Thus, there are thirteen areas which may be divided into eight regions of potentially high tidal-current speeds:

1. North-west Europe.

2. North America and Canada (including (i) north-east USA and Newfoundland, (ii) north-west USA, and (iii) Alaska).

3. Barents Sea.

4. South America (including (i) the Amazon, (ii) the River Plate, and (iii) the Falklands).

5. East Africa.

6. West India.

7. Australasia (including (i) northern Australia and (ii) the east Indies).

8. China Sea and Japan.

Thus, Chapter 6 (North-west Europe) deals with region 1, Chapter 7 (North America and Canada) deals with region 2, and Chapter 8 (Australasia) deals with region 7. Chapter 9 deals with the other areas.

5.3   Tidal datum and tidal heights

Although the details vary around the world's hydrographic agencies, the representation, on a chart or in a sailing pilot, of the height of the tide follows similar conventions throughout. A number of representative levels can be identified with reference to tidal curves at Flat Holm along the Bristol Channel in the South West of England (Figure 5.3) and to Figure 5.4:

- **Highest astronomical tide** (HAT). The height of the highest occurring water level under tidal forcing (i.e. excluding freshwater, meteorological, and storm surges).

- **Mean high water Springs** (MHWS). The height of an average Spring tide high water level.

- **Mean high water Neaps** (MHWN). The height of an average Neap tide high water level.

- **Mean sea level** (MSL). The mean water depth at the site.

- **Mean low water Neaps** (MLWN). The height of an average Neap tide low water level.

- **Mean low water Springs** (MLWS). The height of an average Spring tide low water level.

- **Lowest astronomical tide** (LAT). The height of the lowest occurring water level under tidal forcing (i.e. excluding freshwater, meteorological, and storm surges). LAT usually corresponds to the chart datum (CD) on hydrographic charts.

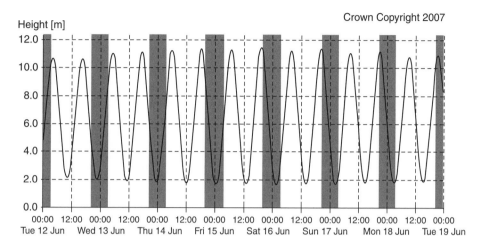

Figure 5.3 Tidal heights at Flat Holm in the Bristol Channel for the period 12–19 June 2007 (from Easy Tide, 2007).

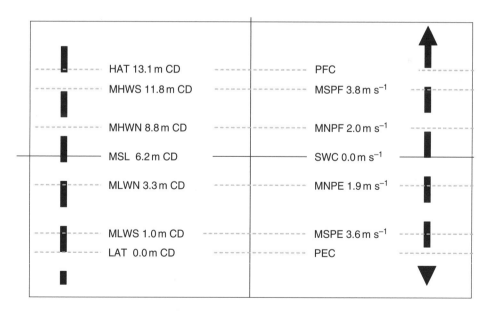

Figure 5.4 Tidal heights on a vertical scale (left) and currents on a speed scale (right). The height data are for Flat Holm in the Bristol Channel, and the currents are for the nearby tidal diamond.

There is then the local topographic datum to which maps are referenced. This would usually correspond to mean sea level, so that hydrographic charts display the difference between chart datum and topographic datum. In the United Kingdom, the topographic datum is called the ordnance survey datum Newlyn, and is referenced to mean sea level at Newlyn in Cornwall (Pugh, 2006).

For the present purposes, we define a corresponding set of parameters to describe the tidal currents at a site (Figure 5.4):

- **Peak flood current** (PFC). The speed of the fastest occurring flood current under tidal forcing (i.e. excluding freshwater, meteorological, and storm surges).

- **Mean Spring peak flood** (MSPF). The speed of an average Spring tide peak flood current.

- **Mean Neap peak flood** (MNPF). The speed of an average Neap tide peak flood current.

- **Slack water current** (SWC). Usually zero.

- **Mean Neap peak ebb** (MNPE). The speed of an average Neap tide peak ebb.

- **Mean Spring peak ebb** (MSPE). The speed of an average Spring tide peak ebb current.

- **Peak ebb current** (PEC). The speed of the fastest occurring ebb current under tidal forcing (i.e. excluding freshwater, meteorological, and storm surges).

In North America, the National Ocean Survey (NOS) is responsible for tidal measurements and for the definition of tidal datums. In San Francisco, for example, the average height of the higher of the two daily high tides is called local mean higher high water (MHHW) (CERES, 2007). The average of all the high tides is called local mean high water (MHW). There are many other datums, including mean lower low water (MLLW), mean low water (MLW), and mean tide level (MTL), which is midway between MHW and MLW. Mean sea level (MSL) is the average of all the tide measurements. Local MLLW is the zero datum of the tides. Values for mean sea level in 1929 were adopted as the national geodetic vertical datum (NGVD 29), or zero elevation for measuring land height. Benchmarks were established throughout the United States, marking local elevations relative to NGVD 29. Since then, disturbance and loss of many benchmarks has warranted a new datum. The NGVD 29 is being replaced by the North American vertical datum of 1988 (NAVD 88), and new geodetic datums are being planned to make use of improved surveying technology.

5.4   British Admiralty tidal diamonds

5.4.1   Determining MSPF, MNPF, SWC, MNPE, and MSPE

British Admiralty charts list the tidal height parameters identified earlier for major ports on each chart. The flow data are displayed in features known as tidal diamonds; there is a reference letter printed at the location of the data on the chart itself, and the Neap and Spring flow speeds and directions are tabulated separately. A similar feature is available on the British Admiralty software package called Total Tide (UKHO, 2004). The Total Tide height display for Flat Holm in the Bristol Channel is shown in Figure 5.5 (left). These data are used as inputs to the STEM resource analysis program as described below.

Figure 5.5 (right) shows the tidal diamond close to Flat Holm in the Bristol Channel. The tide floods in a north-easterly direction into the Severn Estuary until about high water, after which it turns towards the south-west and ebbs back to sea. The maximum Spring and Neap flood currents are 3.8 and 2.0 knots, and the corresponding ebb values are 3.6 and 1.9 knots. These data are also used as inputs to the STEM resource analysis program as described below.

5.4.2   Determining PFC and PEC

Since the peak flows are not given on the British Admiralty tidal diamonds, the correlation between depth and flow has to be determined. Hardisty (1996) demonstrated the correlation from measurements at the Immingham Oil Terminal on the south bank of the Humber Estuary. Data were collected over a Neap–Spring–Neap tidal cycle at Station 1 on the upstream, western end of the jetty (thus exposed to the full ebb, but somewhat sheltered from the flood currents by the jetty). Additional data were collected at Station 2 on the downstream, eastern end of the jetty

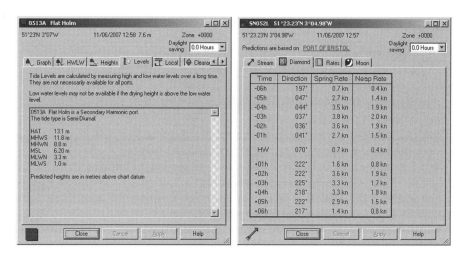

Figure 5.5   (Left) The tidal height data for Flat Holm in the Bristol Channel. (Right) The tidal diamond data for the nearby station (from UKHO, 2004).

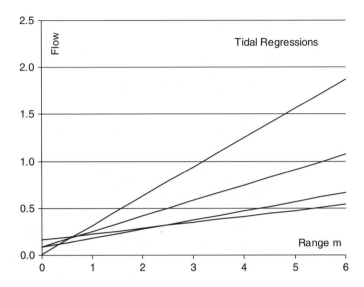

Figure 5.6 Regression equations for the depth–speed correlations at the Immingham Oil Terminal. The data from top to lowest plot apply to Station 1 ebb, Station 2 ebb, Station 2 flood, and Station 2 ebb respectively (author's data).

(exposed to the flood, but sheltered from the ebb). Current speed and direction were measured with Sensordata SD-600 self-recording rotor current meters. The results can be summarized by the regression equations (Figure 5.6):

Station 1 (ebb exposed):

$$PeakEbbFlow = 0.3108R + 0.0070, \ R^2 = 0.05 \tag{5.1}$$

$$PeakFloodFlow = 0.0625R + 0.1632, \ R^2 = 0.78 \tag{5.2}$$

Station 2 (flood exposed):

$$PeakEbbFlow = 0.1638R + 0.0907, \ R^2 = 0.87 \tag{5.3}$$

$$PeakFloodFlow = 0.0970R + 0.0836, \ R^2 = 0.78 \tag{5.4}$$

Thus, the regression equation can be used to determine the peak flood and peak ebb currents corresponding to the highest astronomical tides.

Table 5.1 shows data from the Total Tide package for the Immingham site. Data for 21 March 21 2007 show an Equinoctial high water of 7.9 CD and an Equinoctial low water of 0.3 CD, thus giving an Equinoctial range of 7.6 m. These figures compare with a mean high water Spring (MHWS) of 7.3.

Alternatively, some simplified algebra can be used. Assume

$$\frac{PFC}{(HAT - MSL)} = \frac{MSPF}{(MHWS - MSL)} = \frac{MNPF}{(MHWN - MSL)} \tag{5.5}$$

Table 5.1 Immingham Station 0173 (UKHO, 2004)

	High water CD	Low water CD
30 March 2006	7.9	0.2
10 September 2006	8.0	0.3
21 March 2007	7.9	0.3
28 September 2007	7.9	0.5
22 March 2008	7.3	0.9
18 September 2008	7.6	0.9

Hence

$$\text{PFC} = 0.5\,(\text{HAT} - \text{MSL}) \left(\frac{\text{MSPF}}{(\text{MHWS} - \text{MSL})} + \frac{\text{MNPF}}{(\text{MHWN} - \text{MSL})} \right) \qquad (5.6)$$

Similarly

$$\frac{\text{PEC}}{\text{MSL}} = \frac{\text{MSPE}}{\text{MSL} - \text{MLWS}} = \frac{\text{MNPE}}{\text{MSL} - \text{MLWN}} \qquad (5.7)$$

Hence

$$\text{PEC} = 0.5\text{MSL} \left(\frac{\text{MSPE}}{\text{MSL} - \text{MLWS}} + \frac{\text{MNPE}}{\text{MSL} - \text{MLWN}} \right) \qquad (5.8)$$

Equations (5.6) and (5.8) can be used to determine the values of the peak currents.

5.5 Harmonic decomposition

5.5.1 Introduction

The resource analysis method employed here is embedded in a computer program that is available to be downloaded from the book's website. The program is called the *simplified tidal economic model*, or STEM. STEM is designed to simulate tidal currents and electrical power output at hourly intervals over a 12 month period, making due allowance for Neap and Spring tides, Equinoctial tides, shallow water, and diurnal effects. STEM utilizes only six tidal species (cf. Table 2.4) which are described by, for example, the Manly Hydraulics Laboratory (2008) as:

- M_2 **Principal lunar semi-diurnal constituent.** This constituent represents the rotation of the Earth with respect to the Moon. Period = 12.4206 h.

- S_2 **Principal solar semi-diurnal constituent.** This constituent represents the rotation of the Earth with respect to the Sun. Period = 12.0000 h.

- M_4 **Shallow water overtides of principal lunar constituent.** A short-period harmonic term introduced into the formula of tidal-current constituents to take account of the change in the form of a tide wave resulting from shallow water conditions. Period = 6.2103 h.

- **K_2 Lunisolar semi-diurnal constituent.** This constituent modulates the amplitude and frequency of M_2 and S_2 for the declination effect of the Moon and Sun respectively. The Australian Hydrographic Service (2007) usefully defines the low water Equinoctial Springs that occur near the time of equinoxes as a tidal elevation depressed below mean sea level by the amount equal to the sum of amplitudes of the constituents M_2, S_2, and K_2. Period = 11.967 h.

- **K_1 Lunisolar diurnal constituent.** Period 23.930 h.

- **O_1 Principal lunar diurnal constituent.** Period 25.820 h.

For the present purposes, we signify the amplitude of the flow (as opposed to the height) of the species by the addition of a 'U' to the subscript. Thus M_{U2} is the flow amplitude corresponding to M_2. We may use the Admiralty tidal diamond tidal-current speeds to provide estimates of the amplitudes of the components. For the present purposes, we take a 'typical' 12 month period initiated at phase zero (all of the harmonics are in phase).

5.5.2 Formzahl number

Defant (1961, p. 306) explains that, in order to classify the tides at a locality, four principal types of tide based upon the ratio of the sum of the amplitudes of the diurnal components ($K_1 + O_1$) to the sum of the semi-diurnal components ($M_2 + S_2$) may be considered. This ratio increases when the diurnal inequality of the tides increases. It attains a maximum when there is only one high water in each day. The ratio is called the Formzahl number, Fz:

$$Fz = \frac{K_1 + O_1}{M_2 + S_2} \tag{5.9}$$

where $Fz < 0.25$ is semi-diurnal, $0.25 < Fz < 1.5$ is mixed semi-diurnal dominant, $1.5 < Fz < 3.0$ is mixed diurnal dominant, and $Fz > 3.0$ is diurnal, as described below.

Figure 5.7 shows the distribution of semi-diurnal, diurnal, and mixed tides around the waters of the world's oceans:

(i) Diurnal tides are generated in north-west Canada, the Gulf of Mexico, the Antarctic, south-west Australia, east Java and Thailand, the Sea of Japan, the Sea of Okhotsk, and the Kamchatka Basin.

(ii) Mixed tides are generated on the west coast of North America, the east coast of Central America, Cuba, and west Florida, Greece, the west coast of Madagascar, the Indian Ocean coasts from Ethiopia to the southern extremity of India, Indonesia (except for the diurnal coasts identified above), parts of west, south, and north Australia, Japan, and along the north-west shoreline of the Sea of Okhotsk.

(iii) All other coastlines exhibit semi-diurnal tides.

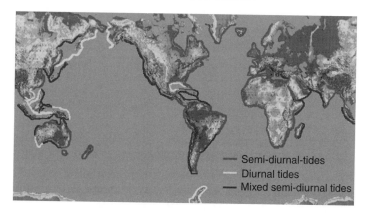

Figure 5.7 Distribution of semi-diurnal, diurnal, and mixed tides around the shores of the world's oceans (from NOAA, 2008). A URL link to the full colour original of this image is available at the book's website.

Thus, the Formzahl number will take a value of <0.25 for the coastlines in (i), >3.0 for the coastlines in (iii), and a value between 0.25 and 3.0 for the coastlines in (ii).

5.5.3 The Neap–Spring cycle, M_{U2} and S_{U2}

The Neap–Spring cycle results from the addition of the first two current speed harmonics, the amplitudes of which will be designated M_{U2} for the lunar semi-diurnal constituent and S_{U2} for the solar semi-diurnal constituent.

Thus, in Figure 5.8 (left), we consider first the maximum flood currents. Let the amplitudes of the M_{U2} be $1.00\,\mathrm{m\,s^{-1}}$, which is added to the S_{U2} amplitude of $0.50\,\mathrm{m\,s^{-1}}$, giving a tide that, when in phase at Spring tides, has a current speed amplitude, called the mean Spring peak flood (MSPF), of

$$\mathrm{MSPF} = M_{U2} + S_{U2} = 1.5 \tag{5.10}$$

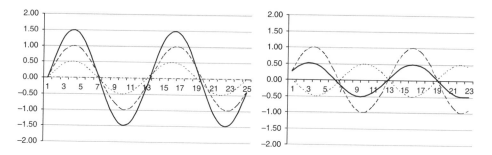

Figure 5.8 (Left) Addition of the M_{U2} (dashed line) and S_{U2} (dotted line) harmonics in phase at a Spring tide, demonstrating equation (5.9). (Right) Addition of the same amplitude harmonics when out of phase at a Neap tide.

or, when out of phase at Neap tides, has a current speed amplitude, called the mean Neap peak flow (MNPF), of

$$\text{MNPF} = M_{U2} - S_{U2} = 0.5 \tag{5.11}$$

Equations (5.10) and (5.11) may be combined. For example, equation (5.10) becomes

$$S_{U2} = \text{MSPF} - M_{U2} \tag{5.12}$$

Hence, rearranging equation (5.11), and substituting with equation (5.12), yields

$$M_{U2} = \frac{\text{MSPF} + \text{MNPF}}{2} \tag{5.13}$$

Substitution of M_{U2} from equation (5.13) into equation (5.12) yields

$$S_{U2} = \frac{\text{MSPF} - \text{MNPF}}{2} \tag{5.14}$$

Equations (5.13) and (5.14) check for MSPF = 1.5 and MNPF = 0.5, yielding M_{U2} of 1.0 m and S_{U2} of 0.5 m, as initially defined. Thus, an approximate result for the amplitudes of the lunar and solar semi-diurnal harmonics can be determined simply from the tidal diamond flow maxima at mean Spring and Neap tides (the values of MSPF and MNPF), provided that the site is strongly semi-diurnal, which would be confirmed by a Formzahl number of less than 0.25.

5.5.4   Shallow water effects, M_{U4}

We saw in Chapter 3 that the tidal wave changes from exhibiting the properties of a progressive wave in deep water to exhibiting the properties of a reflected standing wave in shallow water, bays, and estuaries. The result is that the surface wave and the tidal currents become asymmetric, with a shorter, sharper flood followed by a weaker, more drawn-out ebb. The effect is achieved by frequency leakage, so that energy from the M_2 lunar semi-diurnal harmonic is transferred into the M_4 lunar quarter-diurnal harmonic (Hardisty, 2007). Although these data apply to the shape of the surface wave, the current speeds will necessarily change their profile to account for the movement of water under the wave. This is achieved by the addition of the M_4 harmonic.

This process may be illustrated with the STEM software. Figure 5.9 (left) shows a simple tide consisting of $M_{U2} = 1.0$ and $S_{U2} = 0.5$, as before. Figure 5.9 (right) shows the effect of the addition of a quarter-diurnal component, in the case illustrated $M_{U4} = 0.3$, producing flow asymmetry as the tide moves into shallow water.

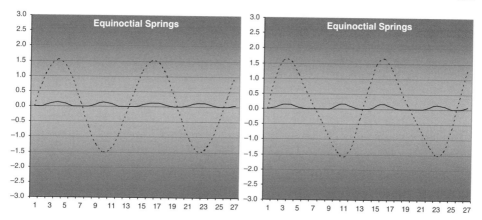

Figure 5.9 (Left) Addition of the M_{U2} and S_{U2} (dotted line) harmonics in phase at a Spring tide. (Right) The same values but with the addition of $M_{U2} = 0.3$, illustrating the increasing asymmetry as the tide moves into shallow water. The solid line represents the STEM power output for a standardized turbine, as described below.

The process is exemplified with data from the Easy Tide package for the Humber Estuary shown in Figure 5.10. The tide progresses up the estuary from Bull Fort at the mouth and moves past Immingham and Kingston upon Hull. The tide becomes more asymmetric and assumes the very distinctive shallow water shape at Blacktoft before it finally reaches Goole, some 90 km from the mouth.

5.5.5   Semi-annual cycle, K_{U2}

The declination is the angle that the Moon or the Sun makes with the equator (Adams, 2005). During the revolution of the Earth around the Sun, the successive positions of the point on the earth which is nearest to the Sun will form a diagonal line across the equator. At the vernal equinox (21 March) the equator is vertically under the Sun, which then declines to the north until the summer solstice (21 June), when it reaches its maximum north declination. It then moves southwards, passing vertically over the equator again at the autumnal equinox (21 September), and reaches its maximum southern declination on the winter solstice (21 December). The declination varies from about 24° above to 24° below the equator. The Sun is nearest to the Southern Ocean, where the tides are generated, when it is in its southern declination, and furthest away when in the north, but the Sun is actually nearest to the Earth on 31 December (perihelion) and furthest away on 1 July (aphelion), the difference between the maximum and minimum distance being one-thirtieth of the whole.

The Moon travels in a similar diagonal direction around the Earth, varying between 18.5° and 28.5° above and below the equator respectively. The change from north to south declination takes place every 14 days, but these changes do not necessarily take place at the change in the phases of the Moon. When the Moon is south of the equator, it is nearer to the Southern Ocean, where the tides are generated.

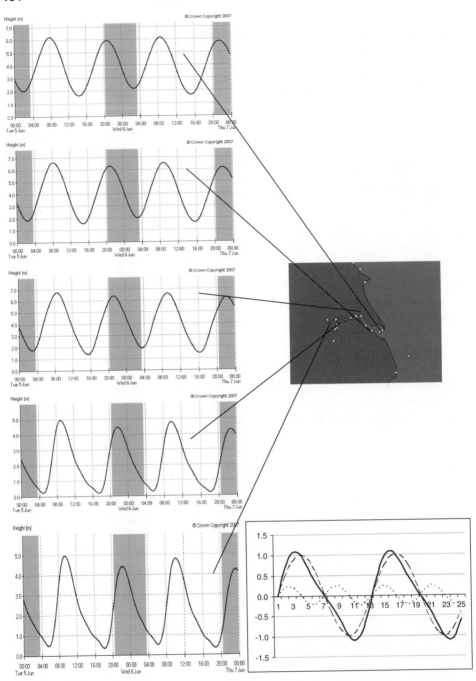

Figure 5.10 The surface tidal wave becomes asymmetric owing to the growth of the M_4 harmonic in an upstream direction, illustrated here for the Humber Estuary from (upper to lower) Bull Fort, Immingham, Kingston upon Hull, Blacktoft, and Goole for 6–7 June 2007. Lower right is a synthetic addition of $M_2 = 1.00$ m and $M_4 = 0.25$ m to simulate the shallow water effects. Humber data from Easy Tide, 2007.

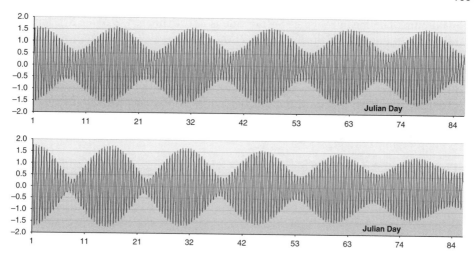

Figure 5.11 Three month (e.g. March to June) current speed simulation (upper) $M_{U2} = 1.00$, $S_{U2} = 0.50$, $M_{U4} = 0.10$, and $K_{U2} = 0.00$ shows no difference in the heights of the maximum Spring tides. However (lower), as above but with $K_{U2} = 0.25$ shows an increase in the maximum speeds of the Spring tides (the Equinoctial tides) from the mean value of 1.50 m to a value of 1.75 m.

The new Moon is nearest to the Sun, and crosses the meridian at midday, while the full Moon crosses it at midnight.

The height of Spring tides varies throughout the year, being at a maximum when the Sun is over the equator at the equinoxes and at a minimum in June at the summer solstice when the Sun is furthest away from the equator (Figure 5.11). Extraordinarily high tides may be expected when the Moon is new or full, and in position nearest to the Earth at the same time as the declination is near the equator. The amplitude of the K_{U2} harmonic accounts for these differences and is simply the difference between the mean Spring and the Equinoctial maximum flood or ebb currents.

5.5.6 Diurnal inequality, K_{U1} and O_{U1}

Defant (1961) gives the following classification:

- **Formzahl number $Fz < 0.25$ semi-diurnal type.** Two high waters and two low waters daily that are of approximately the same height. The mean range at Spring tide is $2(M_2 + S_2)$, and the range on an Equinoctial Spring tide is $2(M_2 + S_2 + K_2)$.

- **Formzahl number $0.25 < Fz < 1.5$ mixed semi-diurnal dominant type.** There are daily two high waters and two low waters, which, however, show inequalities that attain their maximum when the declination of the Moon has reached its maximum. The mean range at Spring tide is approximately $2(M_2 + S_2)$, and the range on an Equinoctial Spring tide is $2(M_2 + S_2 + K_2)$.

- **Formzahl number $1.5 < Fz < 3.0$ mixed diurnal dominant type.** Occasionally, only one high water a day, following maxim declination of the Moon. At other times there are two high waters in the day, which. however, show strong inequalities, especially when the Moon has crossed the equator. The mean range at Spring tide is approximately $2(K_1 + O_1)$.

- **Formzahl number $Fz > 3.0$ diurnal type.** Only one high water daily, and the semi-diurnal almost vanishes. The mean range at Spring tide is approximately $2(K_1 + O_1)$.

Thus, K_{U1} and O_{U1} account for the change from diurnal to semi-diurnal tides. Here, we replace Fz with Fz_U, denoting the flow Formzahl number, where, for the present purposes, $Fz_U = Fz$. The results are again illustrated with the STEM software. Figure 5.12 (left) shows a typical, semi-diurnal tide with $M_{U2} = 1.0$, $S_{U2} = 0.5$, and $Fz_U = 0$. Figure 5.12 (right) shows the same semi-diurnal and diurnal components, but with a Formzahl number of 3.0. It should also be noted that the power output in the former case was 182 MW h year^{-1} and in the latter case 63 MW h year^{-1}, with exactly the same device parameters in both cases. Semi-diurnal tides are thus able to generate significantly more electrical power than is the case with diurnal tides.

All six of these species are incorporated into the STEM model which is available from this book's website and is described in the following sections.

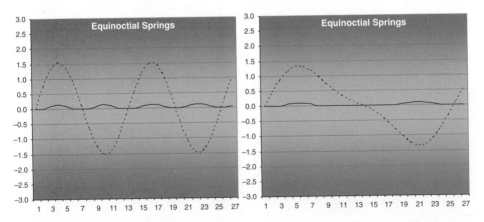

Figure 5.12 (Left) Addition of the semi-diurnal $M_{U2} = 1.0$ and $S_{U2} = 0.5$ harmonics in phase at a Spring tide with a Formzahl number of 0.0. (Right) The same values but with the addition of $Fz_U = 3.0$, illustrating the diurnal result.

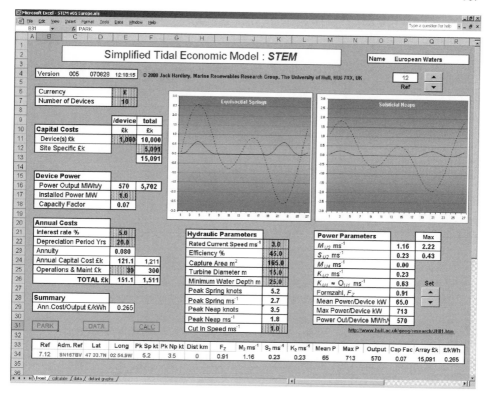

Figure 5.13 Main display and the four sheet tabs.

5.6 STEM display screen

5.6.1 Front display screen

1. The STEM display (Figure 5.13) includes a number of elements. Note firstly the shading conventions:

Red text	data inputs from the data sheet
Black text on shaded cells	constants
Black text on clear cells	calculated results

2. The display includes two front-page input buttons for the user:

Ref	steps through the available data
Set	allows the user to set a non-zero value for the diurnal K_1 and O_1 components

3. There are two display charts showing in solid lines the output power and in broken lines the simulated tidal-current speed throughout 27 h for an Equinoctial tide (left) and Solsticial tide (right).

4. The economic and financial data are shown down the left-hand side of the front-page screen display and are dealt with in the following chapter.

5. The hydraulic parameters are shown in the middle table, which also allows some of the model parameters to be varied as detailed below.

6. The harmonics and power parameters are shown in the right-hand table.

7. The raw data (in red) and calculated results (in black) are shown along the horizontal table towards the bottom of the display.

5.6.2 Tidal and device inputs

There are three sets of inputs for the modelling work:

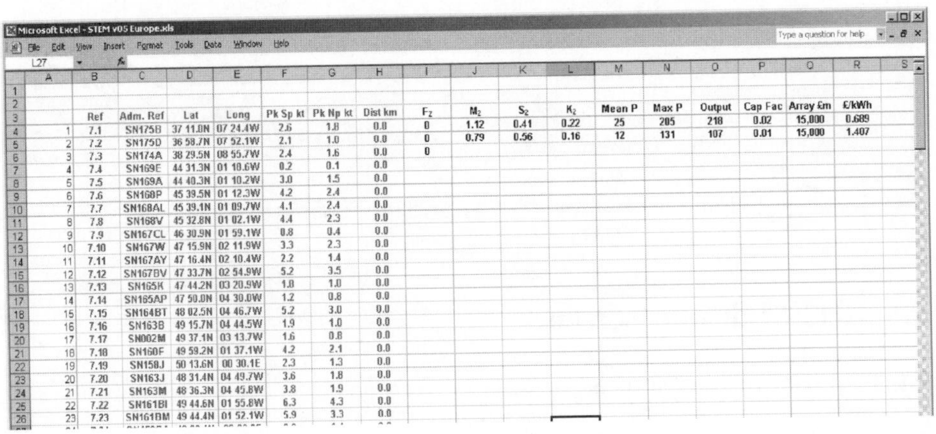

	A	B	C	D	E	F	G	H	I	J	K	L	M	N	O	P	Q	R
1																		
2																		
3		Ref	Adm. Ref	Lat	Long	Pk Sp kt	Pk Np kt	Dist km	F_Z	M_2	S_2	K_2	Mean P	Max P	Output	Cap Fac	Array £m	£/kWh
4	1	7.1	SN175B	37 11.0N	07 24.4W	2.6	1.8	0.0	0	1.12	0.41	0.22	25	205	218	0.02	15,000	0.689
5	2	7.2	SN175D	36 58.7N	07 52.1W	2.1	1.0	0.0	0	0.79	0.56	0.16	12	131	107	0.01	15,000	1.407
6	3	7.3	SN174A	38 29.5N	08 55.7W	2.4	1.6	0.0	0									
7	4	7.4	SN169E	44 31.3N	01 10.6W	0.2	0.1	0.0										
8	5	7.5	SN169A	44 40.3N	01 10.2W	3.0	1.5	0.0										
9	6	7.6	SN168P	45 39.5N	01 12.3W	4.2	2.4	0.0										
10	7	7.7	SN168AL	45 39.1N	01 09.7W	4.1	2.4	0.0										
11	8	7.8	SN168V	45 32.8N	01 02.1W	4.4	2.3	0.0										
12	9	7.9	SN167CL	46 30.9N	01 59.1W	0.8	0.4	0.0										
13	10	7.10	SN167W	47 15.9N	02 11.9W	3.3	2.3	0.0										
14	11	7.11	SN167AY	47 16.4N	02 10.4W	2.2	1.4	0.0										
15	12	7.12	SN167BV	47 33.7N	02 54.9W	5.2	3.5	0.0										
16	13	7.13	SN165K	47 44.2N	03 20.9W	1.0	1.0	0.0										
17	14	7.14	SN165AP	47 50.0N	04 30.0W	1.2	0.8	0.0										
18	15	7.15	SN164BT	48 02.5N	04 46.7W	5.2	3.0	0.0										
19	16	7.16	SN163B	49 15.7N	04 44.5W	1.9	1.0	0.0										
20	17	7.17	SN002M	49 37.1N	03 13.7W	1.6	0.8	0.0										
21	18	7.18	SN160F	49 59.2N	01 37.1W	4.2	2.1	0.0										
22	19	7.19	SN158J	50 13.6N	00 30.1E	2.3	1.3	0.0										
23	20	7.20	SN163J	48 31.4N	04 49.7W	3.6	1.8	0.0										
24	21	7.21	SN163M	48 36.3N	04 45.8W	3.8	1.9	0.0										
25	22	7.22	SN161BI	49 44.6N	01 55.8W	6.3	4.3	0.0										
26	23	7.23	SN161BM	49 44.4N	01 52.1W	5.9	3.3	0.0										

Figure 5.14 Data input sheet.

1. **Data inputs**. The user enters a small number of values and parameters from careful scrutiny of the tidal diamonds or tidal-current tables onto the data sheet, as shown in Figure 5.14. These are described in further detail in the following section:

   (a) STEM reference, e.g. 7.1 in this example;
   (b) Admiralty Reference for the station, e.g. SN175B;
   (c) latitude, e.g. 37 11.0 N;
   (d) longitude, e.g. 07 24.4 W;
   (e) peak Spring and Neap currents (knots), e.g. 2.6 and 1.8;
   (f) Formzahl number (as described below).

2. **Economic parameter inputs.** The user can set up the model for a particular device by typing directly onto the front page of the spreadsheet. These include some economic parameters:

   (a) the currency, e.g. £ sterling;
   (b) the number of devices in the array, e.g. 10;
   (c) the device cost, e.g. £ 1 000 000;
   (d) the site-specific cost, e.g. £ 5 000 000;
   (e) the installed power, e.g. 1 MW;
   (f) the interest rate, e.g. 5 %;
   (g) the depreciation period, e.g. 20 years.
   (h) the operations and maintenance cost/device, e.g. £ 30 000 year^{-1}.

3. **Device parameter inputs.** The user can set up the model for a particular device by typing directly onto the front page of the spreadsheet. These include some device parameters:

   (a) rated current speed, e.g. 3.0 m s^{-1};
   (b) device efficiency, e.g. 45 %;
   (c) capture area, e.g. 165 m^2;
   (d) turbine diameter, e.g. 15 m;
   (e) minimum water depth, e.g. 25 m;
   (f) cut in speed, e.g. 1.0 m s^{-1}.

5.6.3 Hydraulic device inputs

There are six hydraulic parameters that may be set for a particular device:

1. **Rated current speed.** This is the current speed at which the device achieves the installed power output, e.g. 3 m s^{-1}.

2. **Efficiency.** The overall device efficiency was defined in earlier chapters from the relationship between the available hydraulic power and the electrical output:

$$P_E = \frac{1}{2} e\rho A U(t)^3 \tag{5.15}$$

   and is set to 45 % in the present example.

3. **Capture area.** The capture area for the device, e.g. 165 m^2.

4. **Turbine diameter.** The turbine diameter (height) of the device, e.g. 15 m.

5. **Minimum water depth.** This is determined, after, for example, Moore (2007a), to allow the device to be mounted outside the bottom 25 % and at least 5 m below the surface to avoid wave action, e.g. 25 m.

6. **Cut-in speed.** The cut-in speed of the device, e.g. 5 m s^{-1}.

5.6.4 Operation and results

1. STEM reads the mean Spring and Neap peak flows from the data sheet onto the front sheet in knots and converts the results into m s^{-1} (cells K27 to K30).

2. STEM calculates the M_{U2} and S_{U2} amplitudes, on the basis of the assumption that the tide is a pure semi-diurnal tide, from the equations above:

$$M_{U2} = \frac{\text{MSPF} + \text{MNPF}}{2} \tag{5.13}$$

$$S_{U2} = \frac{\text{MSPF} - \text{MNPF}}{2} \tag{5.14}$$

and enters the results into cells Q22 and Q23 respectively.

3. STEM reads the value of the Formzahl number Fz_U and calculates $(K_{U1} + O_{U1})$ by adopting equation (5.9):

$$K_{U1} + O_{U1} = Fz_U (M_{U2} + S_{U2}) \tag{5.16}$$

It is then assumed that $K_{U1} \sim O_{U1}$ (although this can be overwritten by the user), and the result is entered into P26.

4. STEM then uses the specified amplitude of the K_{U1} and O_{U1} components to reduce the values of M_{U2} and S_{U2} from the maxima to zero as a ratio of the Formzahl number to 3, based upon the arguments that tides become fully diurnal at $Fz_U = 3$.

5. STEM takes the value for the amplitude of the lunar quarter-diurnal component, M_{U4}, from cell P24. This is normally set to zero, but can be overwritten by the user.

6. STEM takes the value for the amplitude of the lunisolar semi-diurnal component, K_{U2}, from cell P25. This is normally set to 5 % of the amplitude of the M_{U2}, but can be overwritten by the user.

7. STEM calculates the value for the current Formzahl number Fz_U by adopting equation (5.9):

$$Fz_U = \frac{K_{U1} + O_{U1}}{M_{U2} + S_{U2}} \tag{5.16}$$

8. STEM then calculates the tidal-current speeds in terms of harmonic components due to the solar and lunar interactions described above (e.g. Dyer, 1997;

Hardisty, 2007). The instantaneous horizontal current speed, $U(t)$ (m s^{-1}), was given by

$$U(t) = U_0 + U_{MU2} \cos\left[\frac{2\pi t}{T_{MU2}} + \rho_{MU2}\right] + U_{SU2} \cos\left[\frac{2\pi t}{T_{SU2}} + \rho_{SU2}\right]$$
$$+ U_{MU4} \cos\left[\frac{2\pi t}{T_{MU4}} + \rho_{MU4}\right] + U_{OU1} \cos\left[\frac{2\pi t}{T_{OU1}} + \rho_{OU1}\right] \quad (5.17)$$
$$+ U_{KU1} \cos\left[\frac{2\pi t}{T_{KU1}} + \rho_{KU1}\right] + U_{KU2} \cos\left[\frac{2\pi t}{T_{KU2}} + \rho_{KU2}\right]$$

where A_{MU2}, A_{SU2}, A_{MU4}, A_{KU2}, A_{KU1}, and A_{OU1} are the amplitudes of the lunar semi-diurnal, solar semi-diurnal, lunar quarter-diurnal, lunisolar semi-diurnal, lunisolar diurnal, and lunar diurnal current velocities, T_{MU2}, T_{SU2}, T_{MU4}, T_{KU2}, T_{KU1}, and T_{OU1} are the corresponding periods, taken as 12.420601 h, 12.0000 h, 6.2103 h, 11.967 h, 23.93 h and 25.82 h respectively, ρ_{MU2}, ρ_{SU2}, ρ_{MU4}, ρ_{KU2}, ρ_{KU1}, and ρ_{OU1} are the corresponding phase differences respectively. In practice, all are set to zero at the beginning of the annual simulation in STEM.

9. The software calculates the hydraulic power density per unit area, $P_H(t)$ (W m^{-2}) (equation (2.5)):

$$P_H(t) = \frac{1}{2}\rho U(t)^3 \quad (5.18)$$

where ρ is the density of sea water and $U(t)$ is the current speed.

10. The software determines a mean hydraulic power density, $\overline{P_H}$ (W m^{-2}), over the annual cycle, utilizing hourly values of the instantaneous hydraulic power:

$$\overline{P_H} = \frac{\sum\limits_{t=0}^{t=8772} P_H(t)}{8772} \quad (5.19)$$

11. The software calculates the instantaneous electrical power, P_E (W):

$$P_E(t) = \frac{1}{2}\rho e A U(t)^3 \quad (5.20)$$

12. The software determines and displays the mean device power, $\overline{P_E}$ (W m^{-2}), over the annual cycle, utilizing hourly values of the instantaneous device power:

$$\overline{P_E} = \frac{\sum\limits_{t=0}^{t=8772} P_E(t)}{8772} \quad (5.21)$$

13. The software determines and displays the annual energy output, E_E (kW h), which can be estimated for the 8772 h in a year:

$$E_E = 8772\overline{P_E} \qquad (5.22)$$

5.7   Running STEM

1. Before running STEM, data for a new station should be assembled as shown in Figure 5.14, either from a hydrographic chart or from the British Admiralty Total Tide software package. Identify and copy the appropriate references, depths, and currents into the table (cf. steps 4 and 5 below).

2. The latest version of the software should be downloaded and installed on the user's hard disk. The downloaded copy should not be used (it is useful to set the write block key in the file properties menu), and all work should be done with copies.

3. The software is written in Microsoft Excel, which must be available on the user's machine or network.

4. Open the software and select the 'data' sheet.

5. Select the next available row (or overwrite an existing station).

6. Input data for each column for the new station into columns B to H.

7. Select the front sheet and select the new station number in the STEM Ref input control (upper right). Check that the new station data are displayed in the lower panel.

8. Check and, if required, change the device and flow parameters in the shaded cells.

9. Carefully choose the input harmonic amplitudes for the tidal currents:
   (a) M_{U4}, the lunar quarter-diurnal. Set to zero except in estuarine locations.
   (b) K_{U2}, the semi-annual cycle. Related to the difference between mean and Equinoctial tides.
   (c) The Formzahl number.
   (d) The model is now complete and may, of course, be run for a range of control inputs for further analysis. The results can be copy-pasted onto a blank row on the data sheet for future reference.

5.8   Case study at Flat Holm in the Bristol Channel

The present and following sections utilize the STEM software to compare a coastal site in the English Bristol Channel with an estuarine site in the Humber. The first site chosen for detailed consideration is Flat Holm in the Bristol Channel, south-west

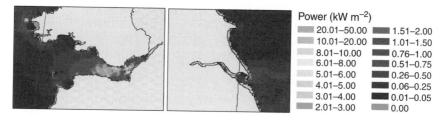

Figure 5.15 Mean annual tidal power for (left) Bristol Channel and (right) west-central North Sea (from DTI, 2004).

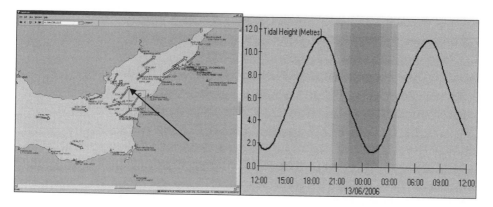

Figure 5.16 Location and tidal depths in the Bristol Channel for 12–13 June 2006 (from UKHO, 2004).

England (Figures 5.15 and 5.16). The approaches to the Bristol Channel from off-shore are mainly a featureless area of sands, although depths increase to nearly 120 m in the gulley known as the Celtic Deep (UKHO, 2002). The Bristol Channel itself has a sandy seabed at its western end, with some mud patches, whereas further east there are areas of gravel and rock outcrops. The northern edge of the channel is characterized by a series of sand banks, the shapes and depths of which are constantly changing.

The tide in the Bristol Channel is predominantly semi-diurnal and increases from a Spring range of about 3.4 m in the west to 12.3 m at the head of the Channel. The maximum tidal range at Avonmouth exceeds 14 m. The heights and ranges of the tide at Flat Holm are given in Table 5.2.

The tidal stream varies considerably through the Bristol Channel. Midway between Hartland Point and Lundy at the entrance to the Bristol Channel, the Spring rate is 3 knots (1.5 m s^{-1}) (UKHO, 2004). Further into the Channel, the rates increase until the entrance to the River Severn is reached, where a rate of 8 knots (4 m s^{-1}) can be attained.

The objective of the following is to define the tidal currents off Flat Holm with the STEM software in order accurately to estimate the available renewable energy resource potential of the site.

Table 5.2 Empirical determination of coefficients in the harmonic analysis and resulting flow and power statistics for the Flat Holm site

Term	Flat Holm	
A_{M2}	1.48	m s^{-1}
A_{S2}	0.46	m s^{-1}
A_{M4}	0.00	m s^{-1}
A_{K2}	0.20	m s^{-1}
Mean current	0.97	m s^{-1}
Mean potential power	0.89	kW m^{-2}
Maximum current	2.14	m s^{-1}
Maximum potential power	5.01	kW m^{-2}
e	0.40	
A	72	m^2
Mean power output	64	kW
Maximum power output	361	kW
Annual energy output	565	MW h
Capacity factor	12.9	%

The data were entered into STEM as described above, and the results are summarized in Table 5.2. The analysis suggests that the mean flow is 0.97 m s^{-1} and the corresponding mean potential power is 0.89 kW m^{-2}. This value is significantly less than the DTI (2004) mean Spring tide value from Figure 5.15. The corresponding maximum flow and potential power density were calculated to be 2.14 m s^{-1} and 5.01 kW m^{-2} respectively. The model was set up with an overall efficiency, e, of 0.40 and a device effective cross-sectional area, A, of 72 m^2 to correspond to the Mark II Neptune Proteus NP1000 device described in Chapter 4. The mean electrical power output was calculated as 64 kW, with a maximum electrical power output of 361 kW and an annual electrical energy output of 565 MW h. The overall capacity factor was determined to be 12.9 %.

5.9 Case study at Hull St Andrews in the Humber Estuary

The other site chosen for detailed consideration is in the Humber Estuary on the east coast of north-central England. The approaches to the Humber Estuary (Figure 5.17) (Hardisty, 2007) from offshore are mainly a featureless area chiefly of sands. The estuary itself has a mainly sandy bottom with finer-grained subtidal and intertidal deposits along the shorelines. The main channel passes seawards through the Humber Gap and then follows the northern shore at Hull Roads and Humber St Andrews before swinging to follow the southern shore in the area of Immingham. The tidal elevations in the Humber are summarized in Table 5.3.

STEM was used to optimize the harmonic coefficients, and the results are shown in Table 5.4. The analysis suggested that the mean flow is 1.44 m s^{-1} and the corresponding mean potential power is 2.63 kW m^{-2}. The corresponding maximum values were calculated to be 2.79 m s^{-1} and 11.09 kW m^{-2} respectively. The model was

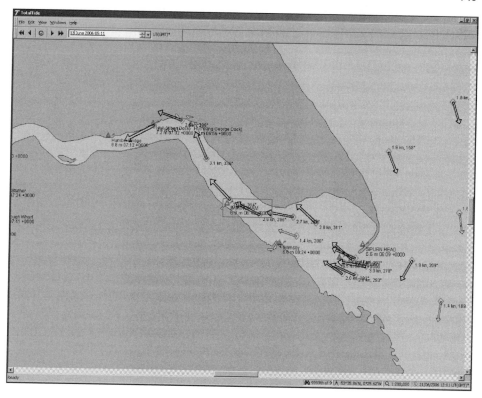

Figure 5.17   The Hull St Andrews site in the Humber Estuary (from HMSO, 2004).

Table 5.3   Principal tidal components for the Humber Estuary

Station	Principal component for the Humber								
	Z_0	M_2	p_{M2}	S_2	p_{S2}	K_1	O_1	F_4	p_{F4}
Spurn Head	4.09	2.13	151	0.72	199	0.16	0.17	0.002	070
Immingham	4.18	2.29	163	0.76	213	0.15	0.18	0.004	196
Hull KG	4.10	2.40	169	0.78	226	0.17	0.16	0.010	270
Bridge	3.59	2.33	174	0.80	228	0.14	0.16	0.026	292
Blacktoft	—	2.05	202	0.60	254	0.11	0.14	0.119	305

again set up for the Mark II Neptune Proteus NP500 with an overall efficiency, e, of 0.40 and a device effective cross-sectional area, A, of 72 m^2. The mean electrical power output was calculated as 181 kW, with a maximum electrical power output of 500 kW and an annual electrical energy output of 1589 MW h year^{-1}. The capacity factor was determined to be 36.2 %.

Table 5.4 Empirical determination of coefficients in the harmonic analysis and resulting flow and power statistics for the Hull Roads site

Term	Hull Roads	
A_{M2}	1.50	m s^{-1}
A_{S2}	0.50	m s^{-1}
A_{M4}	0.05	m s^{-1}
A_{K2}		m s^{-1}
Mean current	1.44	m s^{-1}
Mean potential power	2.63	kW m^{-2}
Maximum current	2.79	m s^{-1}
Maximum potential power	11.09	kW m^{-2}
e	0.40	
A	72	m^2
Mean power output	181	kW
Maximum power output	500	kW
Annual energy output	1589	MW h
Capacity factor	36.2	%

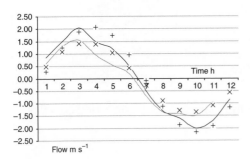

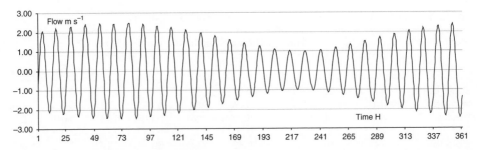

Figure 5.18 Hull St Andrews: (upper) comparison of the Admiralty data for Spring (heavy line) and Neap (light line) with the harmonic model Spring (+) and Neap (×) and (lower) simulation of Spring–Neap–Spring cycle.

5.10 Summary

Although, on a planet with constant water depth with a regular, circular orbit around the Sun and a regular Moon orbit (Chapter 1), the tides would form a simple sinusoidal motion, the complexities of the real orbits and ocean bathymetry result in a number of global amphidromes. The resulting tidal currents reach maximum speeds in eight generalized regions:

1. North-west Europe.

2. North America and Canada (including (i) north-east USA and Newfoundland, (ii) north-west USA, and (iii) Alaska).

3. Barents Sea.

4. South America (including (i) the Amazon, (ii) the River Plate, and (iii) the Falklands).

5. East Africa.

6. West India.

7. Australasia (including (i) northern Australia and (ii) the east Indies).

8. China Sea and Japan.

The details of the local flows are modulated by local channels and coastlines.
  A series of representative tidal levels are usually identified at a particular site:

• highest astronomical tide (HAT);

• mean high water Spring (MHWS);

• mean high water Neap (MHWN);

• mean sea level (MSL), the mean water depth at the site;

• mean low water Neap (MLWN);

• mean low water Spring (MLWS);

• lowest astronomical tide (LAT).

  A series of representative tidal flows are usually identified at a particular site:

• peak flood current (PFC);

• mean Spring peak flood (MSPF);

• mean Neap peak flood (MNPF);

• slack water current (SWC);

- mean Neap peak ebb (MNPE);

- mean Spring peak ebb (MSPE);

- peak ebb current (PEC).

Tidal currents may be synthesized by the identification and summation of a series of harmonic constituents, of which six are utilized here (periods in hours):

- M_{U2}, principal lunar semi-diurnal constituent (12.4206);

- S_{U2}, principal solar semi-diurnal constituent (12.0000);

- M_{U4}, shallow water overtide of principal lunar constituent (6.2103 h);

- K_{U2}, lunisolar semi-diurnal constituent (11.967);

- K_{U1}, lunisolar diurnal constituent (23.930 h);

- O_{U1}, principal lunar diurnal constituent (25.820 h).

The Formzahl number represents the transition from semi-diurnal to diurnal tides and is given by

$$Fz_U = \frac{K_1 + O_1}{M_2 + S_2} \tag{5.9}$$

where $Fz < 0.25$ is semi-diurnal, $0.25 < Fz < 1.5$ is mixed semi-diurnal dominant, $1.5 < Fz < 3.0$ is mixed diurnal dominant, and $Fz > 3.0$ is diurnal, as described below. Figure 5.7 shows the global distribution of the semi-diurnal, diurnal, and mixed tides.

The Neap–Spring cycle results from the addition of the first two current speed harmonics, the amplitudes of which are designated M_{U2} for the lunar semi-diurnal constituent and S_{U2} for the solar semi-diurnal constituent:

$$M_{U2} = \frac{\text{MSPF} + \text{MNPF}}{2} \tag{5.13}$$

$$S_{U2} = \frac{\text{MSPF} - \text{MNPF}}{2} \tag{5.14}$$

This chapter describes the development and operation of the simplified tidal economic model (STEM) which, *inter alia*, calculates the mean device power over the annual cycle by utilizing hourly values of the instantaneous device power:

$$\overline{P_E} = \frac{\sum_{t=0}^{t=8772} P_E(t)}{8772} \tag{5.21}$$

The software determines and displays the annual energy output, E_E (kW h), which can be estimated for the 8772 h in a year:

$$E_E = 8772\overline{P_E} \tag{5.22}$$

STEM is then applied to sites in the Bristol Channel and in the Humber Estuary, yielding the results shown in Tables 5.2 and 5.4.

5.11 Bibliography

Adams, H.C. (2005) *The Sewerage of Sea Coast Towns*. http://www.gutenberg.org/etext/7980 [accessed 5 June 2007].

Australian Hydrographic Service (2007) http://www.hydro.gov.au/prodserv/tides/tidal-glossary.htm [accessed 4 June 2007]; http://www.dominiopublico.gov.br/download/texto/gu007980.PDF [accessed 4 June 2007].

CERES, (2007) http://ceres.ca.gov/wetlands/sfbaygoals/docs/goals1998/draft062698/html/chap05.html [accessed 10 June 2007].

Chalmers University (2008) http://www.oso.chalmers.se/~loading/loadingprimer.html [accessed 21 August 2008].

Defant (1961) [typesetters – this is Given in chapter 1]

DTI (2004) Atlas of UK marine renewable energy resources. Technical Report. http://www.dti.gov.uk/energy/sources/renewables/renewables-explained/wind-energy/page27403.html [accessed 26 March 2008].

Dyer, K.R. (1997) *Estuaries: A Physical Introduction*. John Wiley & Sons, Ltd, Chichester, UK, 195 pp.

Easy Tide (2007) http://easytide.ukho.gov.uk/EasyTide/EasyTide/index.aspx

EPRI, 2005. *EPRI Tidal Instream Energy Conversion Feasibi8lity Demonstration Project: Phase I Feasibility Definition Study. Kickoff Briefing*. Accessed 080530 http://oceanenergy.epri.com/attachments/streamenergy/briefings/2005TidalKickoffMeeting-TidalProject-04-20-05.pdf

ESRU (2007) Strathclyde University, Marine Current Energy Scotland. http://www.esru.strath.ac.uk/EandE/Web_sites/03-04/marine/index2.htm [accessed 10 June 2007].

Hardisty, J. (1996) Immingham oil terminal tidal survey. Unpublished University of Hull Report 96_APT.

Hardisty, J. (2007) *Estuaries: Monitoring and Modelling the Physical System*. Blackwells, Oxford, UK

Manly Hydraulics Laboratory, 2007. http://www.mhl.nsw.gov.au/www/welcome.html Accessed 14 November 2008.

Moore, J., 2007 (a). Quantification of exploitable tidal energy resources in UK waters. Report R10439 ABPMer, Southampton. Text Accessed at http://www.abpmer.co.uk/allnews1623.asp 070722.

Moore, J., 2007 (b). Quantification of exploitable tidal energy resources in UK waters. Report R10439 ABPMer, Southampton. GIS Figures Accessed at http://www.abpmer.co.uk/allnews1623.asp 070722.

NOAA (2008) *Types and Causes of Tidal Cycles – Diurnal, Semidiurnal, Mixed Semidiurnal; Continental Interference*. http://oceanservice.noaa.gov/education/kits/tides/media/supp_tide07B.html [accessed 5 June 2008].

OpenHydro (2008) http://www.openhydro.com/envResource.html [accessed 21 August 2008].

Pugh, D., 1996. *Tides, surges and mean sea level*. John Wiley. Chichester

UKHO (2002) *Admiralty Sailing Directions West Coasts of England and Wales Pilot*, 15th edition. UKHO, Taunton, UK, 383 pp.

UKHO (2004) *Total Tide Tide Prediction Programme*. DP5. UKHO, Taunton, UK.

Part II: Practice

6

Economics and finance

6.1 Introduction

This chapter describes the structure of the renewable energy business in general and, in particular, the economics and finance of tidal stream power. There are presently tens, if not hundreds, of individuals, companies, and organizations involved in any renewable energy development. However, there are also new arrangements and organizations emerging that are taking forward the tidal stream power industry. These are the device and site developers (DSDs). It is the DSDs who are coordinating all of the activities, arranging for the design and manufacture of the hardware, commissioning the system, and raising the vast amounts of capital involved while planning to generate returns on the finance. The business is outlined in Figure 6.1 and described in Section 6.2. The various economic costs and incomes are identified and modelled.

It is, perhaps, a little premature to compile a description of the economics of tidal stream power devices, for very few if any systems have yet been installed on a fully commercial basis. However, many of the details have emerged through various reports (LOG+1, Alstom Power Ltd, & WUTEA, 2007; Siddiqui, Bedard, & Previsic, 2005), and useful parallels may be drawn with the onshore and offshore wind industries (Wizelius, 2007).

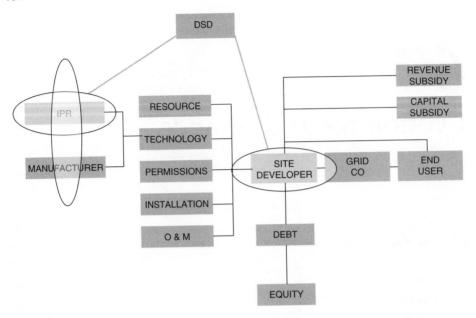

Figure 6.1 The renewable energy business, showing the emergence of the new operators, called the device and site developers (DSDs).

There are a number of schemes available for the identification of economic and financial components of the renewable energy business. Here, we modify the 32 components identified by Siddiqui, Bedard, and Previsic (2005) and the eight comparable components identified in an analysis of wind power by Wizelius (2007) to present the list in four groups:

A. **Device costs.** The device costs include the design and intellectual property (IP) costs and the manufacturing costs detailed below.

B. **Site-specific costs.** The site-specific costs include the various installation permissions and leases, the installation itself, and the costs associated with power transformation and grid connection. Ultimately there will also be site clearance and equipment disposal costs.

C. **Annual costs.** The annual costs include routine operation and maintenance of the device, any unscheduled repairs, as well as the insurances and organizational costs.

D. **Cost of finance.** The cost of finance includes the capital costs of the device and is usually met from loans and from equity investors. The figures need to cover both repayment of loans with interest and the premiums for the investors.

6.2   The Renewable Energy Business

The structure of the renewable energy business is shown in Figure 6.1. From the point of view of the DSDs there are, perhaps, 10 major tasks involved in the development and commissioning of a successful tidal stream power project:

1. **Resource**. This includes the identification of the sites with the required combination of powerful tidal streams, necessary depths and seabed conditions, and access to good grid connections and to maintenance facilities. In particular, a current meter survey campaign may be mounted, a high-resolution numerical model is run, or the kind of harmonic analysis detailed in Chapter 5 is completed in order to describe the tidal current speed and boundary layer conditions.

2. **Technology**. The tidal stream device is chosen from the various designs which are listed in Chapter 4. At this early stage in the development of the industry, it is likely that this decision will be determined as much from product availability as from any proper analysis of the optimal system. The technologists who supply the device will have to have arrangements with two further subgroups:

   (a) **Intellectual property rights**. The designer of the device may well be a separate entity, as appears to be the case with the OpenHydro technology described earlier, or, for example, with Lunar Energy, who have a worldwide, in-perpetuity right to develop and market the RTT turbine. In this case the supplier, or the developer, may need to establish a licensing or other arrangement in order to utilize the design.
   (b) **Manufacturer**. The device is probably manufactured by another company or consortium. In some cases, for example Neptune Renewable Energy, the main company holds the IPR and subcontracts the manufacture of the devices.

3. **Permissions**. Once the site and technology have been identified, there are the processes of obtaining all consents and permissions. The details vary from country to country. In the United Kingdom an Environmental Statement must be prepared and, following consultation with stakeholders, submitted as a formal application for permission from the government agencies.

4. **Installation**. Once the permissions have been obtained and the device has been ordered and delivered, the difficult process of installation can begin. The complexity and cost of the operation depend critically upon the device, and form an important part of the economic analysis detailed below. The work may range from the location of a number of anchors for moored devices such as the Neptune Proteus through to the assembly of gravity foundations, which may involve the lowering of a large containing structure to the seabed to be filled with locally dredged sand and gravel, through to the installation of piles, as for the OpenHydro test device at the European Marine Energy Centre in Orkneys (Figure 6.2).

Figure 6.2 Using a jack-up barge to install the piling works for the OpenHydro device at EMEC, the Orkney Islands (from OpenHydro, 2008).

Installation includes the laying of cables and the establishment of onshore or offshore transformers and the work associated with the actual grid connection.

5. **Operations and maintenance.** Once installed, the systems will require ongoing and long-term management and maintenance. Operations will include maintaining a constant check on the various systems' variables; this is usually done via a communications link to the operator's office, where software displays the various engineering parameters, such as rotation rate, torques, and power delivery, as well as the electricity production to the grid. There will also be a maintenance schedule which may be undertaken under contract and possibly guarantee by the manufacturer for a number of years after installation. Details of the costing of the O&M work are described below.

6. **Debt.** The remainder of this chapter is devoted to a thorough analysis of the economics of tidal stream power devices. In essence, the DSD may have some capital available, may be able to borrow from banks or other lenders, and may release equity in the company to bring investors on board. Debt, with banks or other lenders, is difficult at the present time in the tidal stream power business simply because few if any schemes have yet demonstrated the positive cash flow required for servicing the debt and repaying the loan.

7. **Equity.** Either in parallel with the debt finance or as an alternative to it, most of the tidal stream power schemes involve the raising of some form of equity capital in exchange for a shareholding in the operating company and premiums when the schemes become profitable. Equity investors are usually taking a certain degree of risk on unproven schemes and will be seeking a substantial fraction of the shareholding in exchange. This is less the case for more established schemes. Whatever percentages are finally agreed, the developer

appreciates that a small fraction of a scheme that is up and running and generating a profit is worth very considerably more than a large fraction of a pack of documents, a website, and Power Point presentations.

8. **Grid connection.** Arrangements must be made with the power transmission company, and all of the appropriate 'grid codes' defining tolerances and synchronicity must be complied with.

9. **End-user.** The end-user must be identified if, as is frequently the case with smaller-scale installations, the electricity forms part of a local supply agreement. In other cases, the 'end-user' may simply be the grid company.

10. **Subsidies:**

   (a) **Capital subsidy.** Many governments are using financial instruments to promote the introduction of renewable power. Capital subsidies take the form of non-repayable grants. Typically, the Marine Renewables Deployment Fund (MRDF) in the United Kingdom is presently offering capital grants of 25 % of the installation cost of new devices or arrays.
   (b) **Revenue subsidy.** Again, many governments are offering 'feed-in' payments as a function of the quantity of renewable energy that is generated in order to meet targets. In the United Kingdom, a system of renewable obligation certificates has been established, which, through an auction, provide revenue for the generators.

6.3 Costs

The costs involved in the design, construction, deployment, and operation of a tidal stream power device comprise the device cost, the one-off site-specific costs, and the annual costs. There have been a number of generalized estimates of the costs involved. For example, the Carbon Trust is quoted by Cheng (2007) as examining marine energy and providing the figures in Table 6.1. Prototype tidal stream devices thus have a similar cost to wave power devices but are predicted to become cheaper than wave power devices once production can be initiated.

The capital costs separate into the device costs and the site-specific costs. The device costs are incurred for each device in the array and are detailed in Section 6.3.1. The site-specific costs are one-off costs for each array and are detailed in Section 6.3.2.

Table 6.1 Estimated prototype and production capital costs for wave and tidal stream power devices (after Cheng, 2007)

	Wave Power £ MW[1]	Tidal Power £ MW[1]
Prototype	4.0–8.0	4.8–8.0
Production	1.7–4.3	1.4–3.0

6.3.1 Device costs

(a) **Turbine.** All components that are directly responsible for the extraction of energy from the tidal flow, such as the rotor and its associated controls and the shaft and bearings.

(b) **Extractor structure.** All structural components such as ducting and housing, including finishing, navigational, and safety equipment, as required.

(c) **Power transmission.** Converts the slow movement of the extractor turbine or similar device via gearing or hydraulics into rotary power.

(d) **Electrical machinery.** All equipment including the generator and subsidiary electrical systems for converting the rotary motion into electricity at grid frequency and transmission voltage.

(e) **Instrumentation and control.** Sensors and instrumentation for monitoring the performance of the system, and industrial controllers and feedback systems to initiate command and control sequences. To include all costs associated with the transmission of data from the devices to the shore-based monitoring station.

(f) **Foundation/mooring anchoring.** All components required for holding the tidal-flow power conversion device in place.

(g) **Electrical interconnection.** All cables required to interconnect the individual devices to a common interconnection point.

(h) **Delivery and assembly costs.** Expenditure related to the delivery and assembly of all components to a suitable shore-side site, and shore-side testing and commissioning.

6.3.2 One-off site-specific costs

(a) **Detailed design and specification.** Costs associated with the preparation of the detailed design and drawing up of specifications for the subcontractors and manufacturers.

(b) **Grid interconnection.** All switchgear, transmission lines, and infrastructure required to connect from the common interconnection point to a land-based grid interconnection point.

(c) **Cable.** All cabling from the offshore transformers and switchgear to the onshore transformers and grid connection.

(d) **Communication, command, and control.** All equipment and instrumentation required to establish a two-way link from land-based to tidal-channel-based systems for the purposes of communication, command, and control.

(e) **Installation costs.** The costs required to transport the system from its assembly location to its deployment site and complete all interconnections and checkout to the point where the system is ready to begin commissioning.

(f) **Permits and permissions.** All costs associated with the field survey and formulation of an environmental statement, and with preparing the applications for all necessary permits and permissions, leases, and legal work. The requirements for obtaining a license to install marine energy projects vary. In this section, we give examples from the United States and the United Kingdom.

- **United States.** The US Federal Energy Regulatory Committee (FERC), which regulates the American energy industry, is responsible for granting marine energy licenses and waivers. Given the relative newness of the marine energy space, FERC currently relies on the requirements for established industries such as hydropower when evaluating licensing requests. There are problems with this approach because, for example, the evaluations require data, such as environmental assessments, that can only be obtained through the demonstration projects. As such, on a case-by-case basis, FERC issues preliminary permits that allow companies to conduct demonstration projects in order to gather the required data. Crucially, these demonstration projects are not able to generate electricity for commercial purposes. Upon completion of the demonstration projects, companies may apply for permanent licenses. However, it should be noted that the award of a preliminary permit does not ensure a permanent license. FERC is currently evaluating its application requirements and evaluation process. The aim is to encourage competition and avoid site-banking. Since 2005, FERC has awarded permits for assessment projects in Florida, California, New York, and Washington. They currently (2007) have 45 applications pending.
- **United Kingdom.** There does not appear to be one single agency issuing the licenses, but rather it is a twofold process. The Department of Trade and Industry (DTI) has consent requirements for small-scale demonstration plants, including fulfilling the requirements of the Environmental Impact Assessment and the Habitats Regulations. Once these have been completed, the companies must apply to the Crown Estate, which owns the entire UK seabed out to the 12 nautical mile territorial limit, for a license to generate energy. While there are currently no large commercial-scale projects, should such a project exist, they must complete a full strategic environmental assessment followed by a competitive round for a Crown Estate site lease.

(g) **Owners' development costs.** Assume 5 % of the device costs and site-specific costs detailed above (Siddiqui, Bedard, & Previsic, 2005):

$$O_{\mathrm{DC}} = 0.05 \left(N \sum_{a}^{b} D_{\mathrm{C}} + \sum_{a}^{g} S_{\mathrm{SC}} \right) \tag{6.1}$$

where O_{DC} = owners' development costs, N = number of devices in the array, D_{C} = device costs, and S_{SC} = site-specific costs.

(h) **Spares provision.** The operator or owner will generally maintain a store of the spares required for routine work. Again, Siddiqui, Bedard, and Previsic (2005) recommends a provision of 5 % of the capital costs:

$$S_P = 0.05^* N \sum_{A_1}^{A_8} D_C \tag{6.2}$$

where A_1 to A_8 are the eight device costs listed in 6.3.3.

(i) **General facilities and engineering.** Engineering cost associated with the planning of a tidal-flow farm and the general facilities required for deploying and operating the power plant. This could include necessary dock modifications, maintenance shops, etc., for the deployment and maintenance of the tidal stream power array.

(j) **Commissioning.** The process of inspection and testing required to turn the systems over from the general contractors to the owner operators.

The total plant cost, T_{PC}, is the total installed and commissioned cost of the plant and includes all of the elements above:

$$T_{PC} = N \sum_{a}^{b} D_C + \sum_{a1}^{k} S_{SC} \tag{6.3}$$

6.3.3 Annual costs

The annual costs are made up of the cost of finance, which is itself made up of the cost of investment calculated from the annuity method and the annual cost of operations and maintenance, usually known as O&M.

The investment will probably have been partly financed by equity investors (who may or may not demand a return on their investment) and by a loan. There will be an annual cost for the loan during the repayment period. The annual cost for capital is calculated by the annuity method. The annuity is the sum of amortization (pay-back of the loan) and the interest where the sum of the amortization and the interest will be constant each year. The annual capital cost, C_A, is calculated by the so-called annuity method (Wizelius, 2007):

$$C_A = aC_I \tag{6.4}$$

where a = annuity and C_I = investment cost:

$$a = \frac{r(1+r)^n}{(1+r)^n - 1} \tag{6.5}$$

where r = interest rate and n = depreciation time (years).

The annual operational and maintenance costs have various elements, which are here converted into US dollars from their original (2006) figures, using, where necessary, $US 1 = 8.5 Swedish Kronor (SEK):

(a) **Service.** A tidal turbine will require regular servicing, just like any other machine. The servicing may form part of the device purchase contract under a 3 or 5 year guarantee, or the owner operator may have inherited a service routine from the subcontractors of the different elements of the device. The service crew will make regular monthly or quarterly checks on the device. As there are presently no tidal turbines in regular servicing, it may be useful to compare with the experience of wind turbines. Wizelius (2007) reports that the service costs for wind turbines are often included in the price, but oil and other materials are not. After that period, the manufacturer or a service company offers a service contract, which (2006 figures) will cost about $US 4000–5000 a year (in Sweden) for a 1 MW machine.

(b) **Insurance.** Once again, we can only estimate the costs by comparison with wind turbines. Wizelius (2007) reports that, during the time of the guarantee, usually 2 years, insurance for fire and public immunity costing about $US 400–600 is needed. When the guarantee runs out, machine insurance is usually added. The total cost of insurance for a 1 MW machine is then about $US 500–600 each year.

(c) **Measurement.** The grid operator takes a fee for measuring the production that is fed into the grid. Wizelius (2007) reports that in Sweden this costs about $US 1000 year^{-1} for a 1 MW turbine.

(d) **Telecommunications.** The maintenance of a telecommunications link to the turbine (either through telephony or an Internet protocol data link) will cost about $US 300 year^{-1} in Sweden.

(e) **Taxes.** There may also be taxes and fees to be paid. Wizelius (2007) reports that a 1 MW wind turbine in Sweden will attract a fee to the municipality of about $US 150 year^{-1} and a property fee of $US 4.5 kW^{-1}.

(f) **Administration.** To own and run a turbine also requires a certain amount of administrative work: invoices have to be raised and paid, as do appropriate taxes, and the bookkeeping has to be taken care of. The operation of the turbine needs to be regularly monitored, and routine and exceptional operation and maintenance must be organized. Wizelius (2007) suggests that each 1 MW wind turbine requires at least $US 700 year^{-1} for administrative costs.

6.4 Revenue

The calculation of the income comprises a determination of the power output and the sale price of the electricity, and the identification of any available subsidies. The determination of the power output involves some combination of three techniques: long-term current metering, the development and analysis of high-resolution computer models, or the use of the harmonic analysis method described earlier in this book. The sale price will be the result of a negotiation either with a utility or with a local user. The subsidies that are presently available (2007) in different countries are briefly described in the following sections.

6.4.1 Power output

(a) **Current metering.** In order accurately to determine the power output from the tidal stream device, it is necessary to acquire or to synthesize a long-term current speed and direction time series in order to account for the variability in the resource owing to semi-diurnal, diurnal, Spring–Neap, Equinoctial, and longer-term variations and to the local seabed type and resulting velocity profile. These processes are described and explained in earlier chapters of this book. One method is to commission or to undertake a long-term current metering campaign actually to measure the flows. Although there are a wide range of current meters available on the market at the present time (Hardisty, 2007), the best approach is the deployment of a seabed, upwards-directed, acoustic Doppler current profiler (ADCP), as shown in Figure 6.3.

The ADCP consists of an array of upwards directed acoustic transducers usually operating in the range of, for example, the RDI Workhorse at 300 kHz and a repetition rate of 1–2 Hz. The acoustic signal reflects back off particles in the flow and is range gated so that the signal from different depths is separated. If the particles are moving with the flow, then there is a Doppler shift in the reflected signal which is then converted into a flow speed by the onboard electronics. Thus, a velocity profile from the seabed to the surface is obtained. Data are usually averaged and stored at 20 min or 1 h intervals. The data can then be used as input into a systems model which does not generate electricity below threshold but maximises output

Figure 6.3 Deployment of bottom-mounted ADCP in Juan de Fuca (from UVIC, 2008).

when the installed capacity is achieved. The model also simulates the device's control responses, and a power production curve which is related to current velocity. The instantaneous power outputs are then integrated over 6 months or a year to provide the overall power output.

(b) **Numerical models.** An alternative that is beginning to provide, at least on a regional scale, an estimate of the tidal current streams is the group of numerical tidal stream models that are being developed by national oceanographic institutes and some university groups. For example, Figure 10.4 shows the Proudman Oceanographic Laboratory's continental-shelf-scale maximum Spring tidal current map for the United Kingdom. The map allows areas of interest to be identified but is rarely sufficiently detailed at the local scale to permit accurate power calculations. Another example is shown in Figure 10.5, which is produced by the Australian CSIRO. Both regions will be analysed in more detail in later chapters.

(c) **Harmonic method.** The harmonic method described in Chapter 5 can be used to synthesize long-term flow estimates from which power output is calculated. The analysis can be based upon as little as the peak flood and ebb flows at Neap and Spring tides. It is generally preferable, however, to subject longer-term current meter data to harmonic analysis in order to synthesize more accurate time series.

6.4.2   Sale price

(a) **Delivery contract.** The costs associated with the transportation of the generated electricity from the installation to the user are, usually, charged by the distributor to the generator. There will also be a metering charge for the accurate measurement and recording of the quantity of electricity that is delivered.

(b) **Sale price.** The sale price of the electricity to the distributor or directly to the end-user will depend upon a variety of factors including the rate at which the electricity is delivered, the local, regional, and national demand, and the price of competing sources.

6.4.3   Subsidies

The last 7 years has seen a renewed interest in marine energy as renewable energy becomes more prevalent. While there has been a significant increase in legislative and government support for renewable energy, the majority of this support is focused on solar, wind, hydro, biomass, and geothermal energy. In recent years, this trend has begun to shift to include marine energy. The continued development and commercial viability of marine energy is requiring national, regional, and local governments to introduce investment incentives, tax credits, and obligations that require a percentage of electricity to be derived from renewable sources. Table 6.2 is a summary of several national policies and incentives. Examples from a range of countries are described below.

Table 6.2 National policies supporting marine energy (after Cheng, 2007)

	Germany	Portugal	UK	USA
Capital grants		×	×	
Tax credits	×			×
Obligations			×	
Tradeable certificates			×	
Feed in Tariffs	×	×	×	

(a) **United Kingdom.** The government's main policy aimed at stimulating renewable energy was introduced in 2002 as the Renewables Obligation (Foxon *et al.*, 2005). This policy, which includes marine energy, obligates electricity suppliers to source an increasing percentage of their electricity from renewable sources. The level of obligation rises annually from 6.7 % in 2006/2007 to 15.4 % in 2015/2016. To confirm compliance, the supplier is awarded a Renewable Obligation Certificate (ROC) for each MWh of electricity that is produced from a renewable source. ROCs are tradeable, so that suppliers with a deficit can buy from those with a surplus or from the smaller independent generators. The 2006 price for a single ROC was £47–50. In addition, during 2006, the British government's Department of Transport and Industry launched the Marine Renewables Deployment Fund (MRDF) with the aim of funding the gap between R&D and precommercial deployment of marine energy in the UK. MRDF has up to £50 million allocated for its use. In February 2007, the Scottish Executive announced its intention to award over £13 million in grants for the development of marine energy. The UK government also provides funding to an independent company, the Carbon Trust (CT), whose objective is to assist in moving the country to a low-carbon economy. In 2004–2005 the CT initiated the Marine Energy Challenge, an 18 month initiative designed to accelerate eight wave and tidal programmes in the UK. Grants totalling £3 million were awarded.

(b) **Portugal.** In 2001, Portugal established the National Strategy for Energy, which set a target of 9680 MW of electricity to be generated by renewable energy; 50 MW of this was allocated to wave energy. In 2006, a budget of over €30 million was appropriated for renewable energy sources, €5 million of which has benefited, among other projects, the Pelamis wave farm.

(c) **United States.** In 2005, the USA enacted the Energy Policy Act, which provides tax incentives and loan guarantees for energy production. More importantly, the Energy Policy Act acknowledges marine energy for the first time as a separate renewable energy source qualifying investors for the 10 year Production Tax Credit (Siddiqui, Bedard, & Previsic, 2005). Furthermore, appropriated funds dedicated to renewable energy projects have become available to marine energy projects. In May 2007, the Marine Renewable Energy Research and Development Act was introduced to the House of Representatives. The Act directs the Secretary of Energy to support the research, development, demonstration, and commercial application of marine renewable energy technologies. The Act authorizes appropriations of $US 50 million for each of the fiscal years 2008 through 2012. A renewable portfolio standard (RPS) is a state-by-state requirement that a minimum percentage of each electricity generator's or supplier's resource portfolio derive from renewable energy. An RPS creates a minimum commitment to a sustainable energy future for a given state. It builds on and enhances the investment already made in sustainable energy, and it ensures that new electricity markets recognize that clean renewable electricity is worth more that polluting fossil fuel and nuclear electricity. Further, these goals can be accomplished using a market approach that provides the greatest amount of clean power for the lowest price and an ongoing incentive to drive down costs. By using tradeable 'renewable energy credits' (RECs) to achieve compliance at the lowest cost, the RPS would function much like the Clean Air Act credit-trading system, which permits lower-cost, market-based compliance with air pollution regulations. Renewable portfolio standards (RPSs) exist in 19 states (and may soon exist at a national level) and act as a driver for an REC market. Within the RPSs, a generator of renewable non-polluting electricity can have two income streams: the first from the sale of the electricity generated and the second from the sale of renewable energy certificates that are accrued by generating electrical energy without emissions.

(d) **Canada** provides a production incentive for wind power under the Wind Power Production Incentive (WPPI). The federal government announced a new programme in the 2005 budget – the Renewable Power Production Incentive. There is also an expansion of WPPI, and it will provide 1 cent (Cdn) kW h^{-1} for the first 10 years. For projects commissioned after 31 March 2006, this incentive would be 1¢(Cdn) kW h^{-1} for the first 10 years of production, representing, in the estimate of the Canadian government, about half of the estimated cost premium for wind energy in Canada for facilities with good wind sources. There is no similar explicit production incentive for tidal in-stream developments. Canada also provides the Scientific Research and Experimental Development (SR&ED) federal tax incentive programme to encourage Canadian businesses to conduct research and development in Canada that will lead to new, improved, or technologically advanced products or processes. The SR&ED programme is the largest single source of federal government support for industrial research and development. Applicants may claim SR&ED investment tax credits for expenditures such as wages, materials, machinery, equipment, some overheads, and SR&ED contracts. Generally, a Canadian-controlled private corporation can earn an investment tax credit of 35 % up to the first $2 million of qualified expenditures for SR&ED carried out in Canada, and 20 % on any excess amount. Other Canadian corporations, proprietorships, partnerships, and trusts can earn an investment tax credit of 20 % of qualified expenditures for SR&ED carried out in Canada. To qualify for the SR&ED programme, the project must advance the understanding of scientific relations or technologies, address scientific or technological uncertainty, and incorporate a systematic investigation by qualified personnel.

6.4.4 Annual income

The annual income, I_A, is given simply by the product of the annual output power, P_O, and the sum of the electricity price, E, and the incentives, I. Both the power output and the prices should, of course, have the same units of power, such as kW h year^{-1} or MW h yr^{-1}:

$$I_A = P_O (E + I) \tag{6.6}$$

6.5 Economic result

6.5.1 The annuity method

The economic result (Wizelius, 2007) is the same as the annual profit and is calculated from the annuity method:

$$P_A = I_{NA} - C_A - OM_A \tag{6.7}$$

where P_A = annual profit, I_{NA} = annual income, C_A = annual cost of capital, and OM_A = annual cost of operations and maintenance.

6.5.2 Worked example with the annuity method

Wizelius (2007) provides an example from the analogous case of a Swedish wind turbine proposal, which bears repeating here. A 1 MW wind turbine is expected to produce 2 400 000 kW h year^{-1} at a total investment cost, C_I, of SEK 9 750 000. The total price (inclusive of incentives) obtained for the power is SEK 0.50 kW h^{-1}. The annual O&M cost is SEK 127 000. The annual income is then given by equation (6.6) as 2 400 000 × 0.50 = 1 200 000. The annual net income is then given by 1 200 000 − 127 000 = SEK 1 073 000. If the interest rate is 6 % and the depreciation time is 20 years, then equation (6.7) yields the annuity, a:

$$a = \frac{0.06\,(1 + 0.06)^n}{(1 + 0.06)^{20} - 1} = 0.087185 \tag{6.8}$$

Thus, the annual capital cost, C_A, calculated from the annuity method (equation 6.7), is 0.087185 × C_I = 0.087185 × 9 750 000 = SEK 850 054. The annual profit is then given from equation (6.7) as

$$P_A = I_{NA} - C_A - OM_A = 1\,073\,000 - 850\,054 = SEK\,222\,946 \tag{6.9}$$

The units are Swedish Kronors, where (in January 2006) the exchange rates are given by Wizelius (2007) as ECU 1 = SEK 9.3, $US 1 = SEK 7.7. Thus, the annual profit from an investment of about $US 1.2 million (approximately £600 000) is approximately $US 40 000 (approximately £20 000), or 3 %. Wizelius (2007) confirms that prices in Sweden are very low by international standards.

6.5.3 The present value method

Another method that is employed to calculate the economic result for an investment in tidal turbines is the present value method, which is also known as the discounting method. With this method, the value of an annual income or expense that will occur for a specific number of years is given the value at a specific time, usually the day when the turbine starts to operate. If the present value of revenues is larger than the present value of the investment and expenses, the investment will be profitable. The present value is calculated from

$$PV = f_C \times R \qquad (6.10)$$

where PV = present value, R = revenue (annual net income), and f_C = capitalization factor:

$$f_C = \frac{(1+r)^n - 1}{r(1+r)^n} \qquad (6.11)$$

where r = interest rate and n = depreciation time (years).

6.5.4 Worked example with the present value method

Wizelius provides an example from the analogous case of a wind turbine proposal, which bears repeating here. Assuming an operation time for the turbines of 20 years and a real interest rate of 6 %:

$$f_C = \frac{(1+0.06)^{20} - 1}{0.06(1+0.06)^{20}} = 11.5 \qquad (6.12)$$

The present value of the net income is then given by equation (6.7), which for the revenue (annual net income) from the worked example above is SEK 1 073 000:

$$PV = 11.5 \times 1\,073\,000 = SEK\ 12\,339\,500 \qquad (6.13)$$

The profit during the 20 years of operation will thus be the present value of the net income minus the investment cost:

$$Profit_{20} = 12\,339\,500 - 9\,750\,000 = SEK\ 2\,589\,500 \qquad (6.14)$$

On an average annual basis, this is a return of 2 589 500/20 = SEK 129 000 each year. The units are Swedish Kronors, where (in January 2006) the exchange rates are given by Wizelius (2007) as ECU 1 = SEK 9.3, $US 1 = SEK 7.7. Thus, the annual profit from an investment of about $US 1.2 million (approximately £600 000) is approximately $US 20 000 (approximately £10 000), or 1 %. Recall that Wizelius (2007) confirms that prices in Sweden are very low by international standards.

6.5.5   The pay-back method

A third method used to evaluate the economic preconditions for an investment in a tidal turbine project is the pay-back method. This is used to calculate how long it will take to get back the money that has been invested. The pay-back time, T (years), is calculated with this simple formula:

$$T = \frac{Investment}{AnnualNetIncome} \qquad (6.15)$$

6.5.6   Worked example with the pay-back method

Wizelius provides an example from the analogous case of a wind turbine proposal, which bears repeating here. The pay-back time for the example introduced above with an investment of SEK 9 750 000 and an annual net income of 1 073 000 is simply

$$T = \frac{9\ 750\ 000}{1\ 073\ 000} = 9\ \text{years} \qquad (6.16)$$

6.6   Conclusions

This chapter has reviewed and detailed the various costs and revenue streams involved in the tidal stream power business and began by dividing the business into 10 major components:

(i)   Resource.

(ii)   Technology:

    (a)   intellectual property rights;
    (b)   manufacturer.

(iii)   Permissions.

(iv)   Installation.

(v)   Operations and maintenance.

(vi)   Debt.

(vii)   Equity.

(viii)   Grid Connection.

(ix)   End-user.

(x) Revenue:

  (a) sales;
  (b) capital subsidy;
  (c) revenue subsidy.

The **capital costs** divide into:

(a) The **device costs** including turbine, structure, power transmission, electrical machinery, instrumentation and control, foundation/mooring, electrical, delivery, and assembly costs.

(b) The **one-off site-specific costs** including detailed design, grid connection, cable, communications, installations, permits and permissions, owners' development costs, spares provision, and general facilities and engineering costs.

The **annual income** requires an estimate of the power output, which is obtained from current metering, numerical modelling, and harmonic analysis and divides into:

(a) electricity sales;

(b) subsidies.

The **annual costs** are made up of the cost of finance and the operating and maintenance (O&M) costs. O&M costs may be separated into:

(a) service;

(b) insurance;

(c) measurement;

(d) telecommunications;

(e) taxes;

(f) administration.

The **economic result** can be calculated from:

(a) **The annuity method:**

$$P_A = I_{NA} - C_A - OM_A \qquad (6.7)$$

where P_A = annual profit, I_{NA} = annual income, C_A = annual cost of capital, and OM_A = annual cost of operations and maintenance.

(b) **The present value method:**

$$PV = f_C \times R \tag{6.10}$$

where PV = present value, R = revenue (annual net income), and f_C = capitalization factor:

$$f_C = \frac{(1 + r)^n - 1}{r (1 + r)^n} \tag{6.11}$$

where r = interest rate and n = depreciation time (years).

(c) **The payback method,** where

$$T = \frac{Investment}{AnnualNetIncome} \tag{6.15}$$

6.7 Bibliography

Cheng, P.K. (2007) Green Energy Newsletter, June 2007, Marine Energy.

Foxon, T.J., Gross, R., Chase, A., *et al.* (2005) UK innovation systems for new and renewable energy technologies: drivers, barriers and system failures. *Energy Policy* 33, 2123–2137.

Hardisty, J., 2007. *Estuaries: Monitoring and Monitoring the Physical Systems.* Blackwell, Oxford.

LOG + 1, Alstom Power Ltd, & WUTEA (2007) http://www.dti.gov.uk/files/file37093.pdf accessed 18 November 2008.

OpenHydro (2008) http://www.openhydro.com/news/OpenHydroPR-140107.pdf [accessed 21 August 2008].

Siddiqui, O., Bedard, R., & Previsic, M. (2005) Economic assessment methodology for tidal in-stream power plants. EPRI Report TP-002 NA http://oreg.ca/docs/EPRI/EPRITidalEconMethodology.pdf [accessed 16 June 2007].

UVIC (2008) http://canuck.seos.uvic.ca/rkd/JdF/moorings/adcpmoor.html

Wizelius, T. (2007) *Developing Wind Power Projects: Theory and Practice.* Earthscan, London, UK, 290 pp.

7

North-west Europe

7.1 Introduction

This chapter considers the European waters from the coastlines of Turkey and Greece to Italy, France, and Spain on the northern shores of the Mediterranean Sea. It then focuses on the Atlantic seaboards of Spain, Portugal, France, England, Wales, Ireland, Scotland, and Norway, before turning to the North Sea and the coastlines of Belgium, Holland, Germany, and Sweden (Figure 7.1). From the outset, it is recognized that much of this area can be discounted for the present purposes. Within this chapter, therefore, the Mediterranean is dealt with separately, and only a small number of sites in local, confined straits are considered.

The Analysis of Tidal Stream Power   Jack Hardisty
© 2009 John Wiley & Sons, Ltd

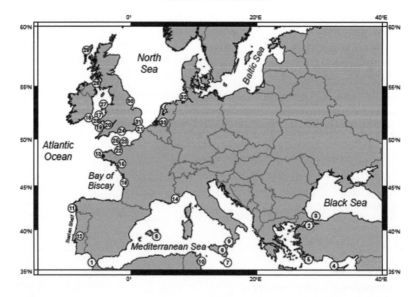

Figure 7.1 North-west Europe. Locations referenced in this chapter are:

1. Straits of Gibraltar	12. River Tagus	23. Cherbourg Peninsula
2. Sea of Marmara	13. Ushant	24. Isle of Wight
3. Bosphorous	14. Cap Ferrat	25. The Alderney Race
4. Cyprus	15. The Loire	26. North Channel
5. Rhodes	16. The Gironde	27. Anglesey
6. Sicily	17. St David's Head	28. Ramsey Island
7. Malta	18. Carnose Point	29. Hebridean Islands
8. Balearic Islands	19. Lundy Island	30. Humber Estuary
9. Straits of Messina	20. Bristol Channel	31. Thames Estuary
10. Gulf of Gabes	21. Strait of Dover	32. Elbe
11. Cape Finisterre	22. St Malo	33. Rhine

The geography and oceanography of the regions are described. The *simplified tidal economic model* (STEM) developed in Chapter 5 is then utilized, alongside some very recent research reported in the literature, to investigate the potential for, and relative cost of, tidal stream power developments in north-west Europe. The conclusion has to be that this area represents vast potential, particularly around the United Kingdom and north-west France, but that developments may, in the short term, be hindered by the costs of the early-stage technology and the installation and grid connection problems.

STEM was set up with 'standardized' inputs for the analyses in this and the following chapters. In addition to the site-specific, locational, and flow parameters, these standardized inputs were introduced in Section 5.6 and are:

- **Data inputs.** A distance from shore of 0 km.

- **Economic parameter inputs.** An array of 10 devices, each costing £1 million, a site-specific cost of £5 million, an installed capacity of 1 MW, an interest rate of 5 %, a depreciation period of 20 years, and an O&M charge of £30 000 year^{-1}.

- **Device parameter inputs.** A rated current speed of $3\,\text{m s}^{-1}$, an efficiency of $45\,\%$, a capture area of $165\,\text{m}^2$, a turbine diameter of $15\,\text{m}$, and a cut-in speed of $1\,\text{m s}^{-1}$.

The STEM outputs are shown as an Appendix at the end of this and the following three chapters.

Nine coastal sites are identified: Pentland and Westray, Islay, Anglesey, Ramsey Island, and the Isle of Wight in United Kingdom waters, the Alderney Race in British and French waters, the Ile de Sein off north-west France, and the Strait of Messina and the Bosphorous in the Mediterranean. In addition, five estuarine sites are identified: the Humber, Severn, and Mersey in United Kingdom waters and the Gironde and Port de Navalo in western France. All of these sites appear to be capable of generating more than $1000\,\text{MW h year}^{-1}$ from the 'standardized' turbine at relative (2008) costs of less than £$0.200\,\text{kW h}^{-1}$.

7.2   Geography of north-west Europe

The geography of north-west European waters can be divided into a series of regions that, proceeding from the Mediterranean into the Atlantic and then northwards from the Strait of Gibraltar and into the North Sea, may be listed as:

   (i) Mediterranean;

  (ii) Iberian Shelf;

 (iii) Bay of Biscay;

 (iv) Celtic Sea;

  (v) English Channel;

 (vi) Irish Sea;

(vii) North-west Approaches;

(viii) North Sea

   (i) **Mediterranean.** The Mediterranean Sea is connected to the Atlantic Ocean by the Strait of Gibraltar in the west and to the Sea of Marmara and Black Sea, by the Dardanelles and the Bosphorus respectively, in the east. The Sea of Marmara is often considered a part of the Mediterranean Sea, whereas the Black Sea is generally considered as a separate entity. The man-made Suez Canal in the south-east connects the Mediterranean Sea with the Red Sea. Large islands in the Mediterranean include Cyprus, Rhodes, and Corfu in the eastern Mediterranean, Sicily and Malta in the central Mediterranean, and the Balearic Islands in the western Mediterranean. The principal tidal power site is the Straits of Messina between Sicily and mainland Italy, although relatively strong currents occur in the Gulf of Gabes.

(ii) **Iberian Shelf.** The Iberian Shelf is a narrow, continental region stretching from a line due west from Gibraltar out to the 200 m isobath, to a line stretching due west from the north-west corner of Spain at Cape Finisterre. The shelf is generally less than 50 km wide off western Spain and Portugal and narrows to only a few kilometres in the south-west. The narrow shelf is dissected by deep canyons running westwards, as, for example, off the mouth of the River Tagus near Lisbon which appears to lead to the Tagus Abyssal Plain. These are almost certainly features eroded or enhanced at times of lower sea level during the Quaternary. The coastal region encompasses southern Spain in the south, Portugal, and then Spain again in the north.

(iii) **Bay of Biscay.** The Bay of Biscay is a much wider shelf area stretching from the northern boundary of the Iberian Shelf, represented by a line due west from Finisterre to the southern boundary of the Celtic Sea, which may be taken as a line running due west from Ushant in Brittany. The French section of the Biscay shelf widens from about 20–50 km in the south to more than 100 km in the north. The shelf is indented by the Cap Ferret Canyon, whereas the two main modern river systems, the Loire and the Gironde, do not seem to be connected to canyon systems. The French section of this shelf is sometimes separated into the Amorican region in the north and Aquitaine in the south (Puillat *et al.*, 2004).

(iv) **Celtic Sea.** The Celtic Sea is the body of continental-shelf water lying south of a line from St David's Head to Carnose Point, and to the west of Lundy Island in the Bristol Channel, and of a line between Land's End and Ushant in the English Channel. It covers an area of some 75 000 km^2 and has four distinct morphological regions:

   (a) **The North-eastern Sill.** This is a relatively flat, featureless seabed with depths between 90 and 100 m, which deepens towards St George's Channel in the north (Belderson & Stride, 1966).

   (b) **Ridges and troughs.** In the south-west there are the elongate, parallel, north-easterly trending ridges and troughs of Great Sole, Cockburn, Jones, and Labadie Banks. These are known as tidal sand ridges (TSRs) and are attributed to tidal controls at a time of lower sea level (Huthnance, 1982; Pantin & Evans, 1984).

   (c) **Shelf edge.** The western boundary of the Celtic Sea is marked by the continental-shelf edge and slope dissected by a series of steep-walled canyons (King Arthur, Whittard, Shamrock, and Black Mud), where the depth increases from about 160 m to the 5000 m of the abyssal plain.

   (d) **Celtic Deep.** In the northern Celtic Sea there is a large, shallow depression, elongated north-north-east-south-south-west and measuring approximately 120 by 30 km, known as the Celtic Deep. It descends to a depth of 20–40 m below the general level of the seabed and is due to continued tectonic subsidence of an older basin. The coastal areas of the Celtic Sea include the high tidal ranges of the Bristol Channel and Severn Estuary, which are discussed further below.

(v) **English Channel.** The English Channel is bounded by the Strait of Dover in the west, the French and English shorelines, and by a line between Cornwall in the north-west and Brittany in the south-east. The seabed is characterized mainly by outcrops of soft, easily eroded Mesozoic, Tertiary, and Quaternary sedimentary rocks and by mobile sediments largely derived from these outcrops. The Channel is shallow and deepens gently from about 30 m in the Strait of Dover to about 70 m in the west. Close to the shore, the seabed profile along the whole Channel may be cliffed or show stepwise changes in gradient. These are related to coastal erosion during periods of lower sea level during the late Tertiary and especially during the Pleistocene glaciations (Cooper, 1948; Wood, 1976; Donovan & Stride, 1975; Hamilton, 1979). The coastal regions of the English Channel include the English estuaries of the Dart, Ex, and Fal, the Solent, and the French estuaries along the coast of Brittany, including the tidal barrage scheme at St Malo. Further to the east, the tidal currents around the Cherbourg Peninsula, the Alderney Race, and the Channel Islands are very considerable and are matched only by the strong flows found to the south of the Isle of Wight. These are discussed in more detail below.

(vi) **Irish Sea.** The general bathymetry of the Irish Sea is described by Bowden (1979). The region comprises the sea area extending from a line joining Carnose Point and St David's Head in the south, to the North Channel between Larne and Corsewall Point. It thus includes St George's Channel (75–140 km wide and 150 km long) between Ireland and Wales and the broader northern area (195 km from east to west and 150 km from north to south) with the Isle of Man in the centre. St George's Channel communicates with the Celtic Sea which is itself open to the Atlantic, while the North Channel is only 20 km wide and connects with the open ocean. The sea has a total area of about 47 000 km². The Irish Sea coast of north-west England includes large areas of sandbanks that are exposed at low tide, particularly the Solway Firth and Morecambe Bay. There are also strong tidal currents off Anglesey and Ramsey Island, which are discussed below. The coastline also includes the estuaries of the Ribble and Mersey. On the east coasts of Ireland and Northern Ireland there are a number of flooded river valleys with strong tidal currents, including Carlingford and Strangford Lough, in the latter of which Marine Current Turbines are installing their *SeaGen* tidal stream power device (Chapter 4).

(vii) **North-western Approaches.** The North-western Approaches form the narrow but deep continental-shelf region between the mainland of Northern Ireland and Scotland and the 200 m isobath. The region can be divided into three morphological areas, with the Malin Sea in the south, the Sea of the Minches, which lies between the Outer Hebrides and mainland Scotland, and the Sea of the Hebrides, which lies outside the islands:

(a) **Malin Sea.** This area lies north of Donegal and Northern Ireland and south of approximately 57 °N. Offshore the seabed descends from about 40 m to some 140 m in a series of broad undulations, while, further south, south-east facing scarps correspond to geological faults and to the boundaries between Tertiary basalts and the nearby Permian to Cretaceous formations (Evans, Whittington, & Dobson, 1986). Locally, tidal currents can be very strong, as discussed below.

(b) **Sea of the Minches.** This area consists of two large interconnected channels, with adjoining lochs and sounds, between the mainland of Scotland and the Hebridean Islands. Bishop and Jones (1979) showed that the most important influence on the bathymetry has been a well-developed preglacial drainage system with fluvial features on the sea floor, first mapped by Ting (1937) and in greater detail by Goddard (1965).

(c) **Sea of the Hebrides.** This area has a bathymetry that takes the form of a broad, asymmetric channel running approximately north-north-east-south-south-west and descending to a maximum depth of 240 m in the west. The channel is subdivided by several banks with deep local depressions, and the banks appear to be due to resistant tertiary bedrock. The western edge of the channel is a well-defined scarp that corresponds to the Minch Fault (Evans, Whittington, & Dobson, 1986).

(viii) **North Sea.** The North Sea is epicontinental and surrounded on three sides by land (Hardisty, 1990). It is open to the Atlantic Ocean in the north, where it continues into the Norwegian Basin, to the Baltic Sea in the east, and to the English Channel to the south through the narrow but important Strait of Dover.

(a) **Northern North Sea.** North of 56 °N, the North Sea evidences depths of 100–140 m. The seabed deepens northwards and away from the coast and evidences narrow deep holes (Swallow, Devils, Swatchway), with maximum depths in excess of 200 m found within an arcuate belt with its apex to the north of Aberdeen that have been attributed to subglacial erosion (Flinn, 1967).

(b) **Southern North Sea.** The southern North Sea and the German Bight are a relatively shallow and featureless area ranging from 30 to 50 m deep, with shallow coastal waters and evidencing an abundance of tidal sandbanks. The coastal regions, particularly of the southern North Sea, include some of north-west Europe's most important estuaries, such as the Humber, the Thames, the Elbe, and the Rhine, and are discussed in more detail below.

7.3   Oceanography of north-west Europe

General circulation. The Mediterranean is largely landlocked, and its circulation is therefore controlled by the fact that evaporation greatly exceeds precipitation and river runoff (Pinet, 1996). Evaporation is especially high in its eastern basin, so that salinity increases eastwards, pushing the relatively cool and low-salinity water from the Atlantic across the basin. The Atlantic water warms and becomes more saline towards the east, before sinking and returning at depth to spill over the Strait of Gibraltar. Thus, sea water flow is eastward in the Strait's surface waters, and westward at depth.

In the Atlantic, however, sea water properties are controlled by the two major circulation patterns known as the *Subpolar Gyre* and the *Subtropical Gyre*, as shown in Figure 7.2. Warm and saline waters flow with the Subtropical Gyre, while colder and less saline waters are transported by the Subpolar Gyre (Hátún *et al.*, 2005).

Temperature. In the Atlantic, the Gulf Stream is represented by warm water travelling northward along the coast of North America and eastward into the central Atlantic Ocean. The result is that the Atlantic coastlines are warmed in winter and cooled in summer around a mean water temperature of about 12 °C, while shallow coastal regions, and particularly the North Sea, cool in winter towards freezing and warm in summer towards about 20 °C.

Salinity. The salinity distribution in the north Atlantic shows two distinct water bodies. In the south are the saline waters associated with the Subtropical Gyre, whereas in the north are the fresh waters associated with the Subpolar Gyre. The result is that the Atlantic coastlines are more saline, whereas shallow coastal regions, and particularly the North Sea, are less saline.

Tidal Elevation. The tidal motion in coastal and continental waters is not the result of the direct action of the gravitational forces of the Moon and the Sun on the waters

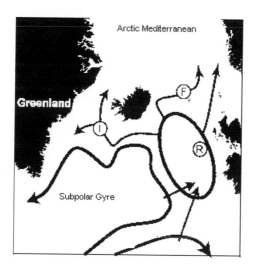

Figure 7.2 North Atlantic circulation patterns are controlled by the Subpolar and the Subtropical Gyres. R is the Rockall Trough, F is the Faroe Current, and I is the Irminger Current.

of the shelf seas. This was shown, for example, by G.I. Taylor, who, in his classical paper of 1918, estimated that the direct attraction of the Moon on the waters of the Irish Sea contributes less than 7 % to the total tidal energy dissipated there during a tidal cycle. The observed tides are, in fact, a response of the shelf seas to the tides generated in the wide expanse of the Atlantic Ocean. The Atlantic tides themselves are not as yet known in detail, but observations and modelling work suggest that a counter-rotation of the tide about an amphidromic point south of Greenland and a similar rotation about a second amphidromic near the Faroe Islands result in a northwards progression of the tide up the western edge of the continental shelf, followed by a swing eastwards across the north of Scotland.

It is conventional (Chapters 1 and 4) (Hardisty, 2008a) to refer to the twice-daily tide due to the Moon as the M_2 tide, and that due to the Sun as the S_2 tide. The progression of the tidal wave across the shelf is best depicted by charts of these tides which show the cotidal and corange lines; cotidal lines join places where high tide occurs simultaneously, while corange lines join places of equal tidal range. Whewell was the first to attempt to draw cotidal lines spanning a sea area, and in 1836 he published a tentative chart for the M_2 in the North Sea (shown, for example, in Proudman & Doodson, 1924). A controversial aspect of Whewell's chart was the suggestion that points of zero tidal range, known as amphidromic points, existed in the open sea, around which the tide revolved. It was not until the beginning of the twentieth century that observations confirmed the existence of amphidromic points, and, in 1924, Proudman and Doodson published the definitive tidal chart for the M_2 tides in the North Sea. Further observations allowed Doodson and Corkan (1932) to extend the work to the English Channels and the Irish Sea and to publish a complete chart of the M_2 corange and cotidal lines for the seas surrounding the British Isles. Lennon (1961) subsequently extended these lines to the edge of the continental shelf (Figure 7.3).

Modern, numerical simulation of the global and regional tidal response has been computed by, for example, the TOPEX team from Oregon State University (Egbert, Bennett, & Foreman, 1994), showing an anticlockwise circulation in the Mediterranean around amphidromic points located, for example, south-west of the Balearics, between Sicily and North Africa and in the Aegean. The amplitudes, however, are not large, peaking at less than 0.5 m.

Marta-Almeida and Dubert (2006) describe a model for the tides on the Iberian Shelf, and the results for the main M_2 and S_2 harmonics show that the tides progress at a regular speed from south to north, and the amplitude increases towards the Bay of Biscay.

Tidal currents. Tidal currents are more difficult to map, since both magnitude and direction of the flow need to be displayed at different times in the tidal cycle. Atlases of tidal streams, therefore, most commonly contain a sequence of charts showing the direction and strength of the tidal currents at hourly intervals during a complete tidal cycle. A different approach, which loses the directional information, is taken by, for example, DTI (2008), where the maximum amplitude of a mean Spring tidal current is displayed. Such an approach is taken by the OSU group (in the work cited), showing, for example, the Mediterranean M_2 tidal currents peaking at a maximum of about 0.1 m s^{-1} in the northern Adriatic and in the central Mediterranean.

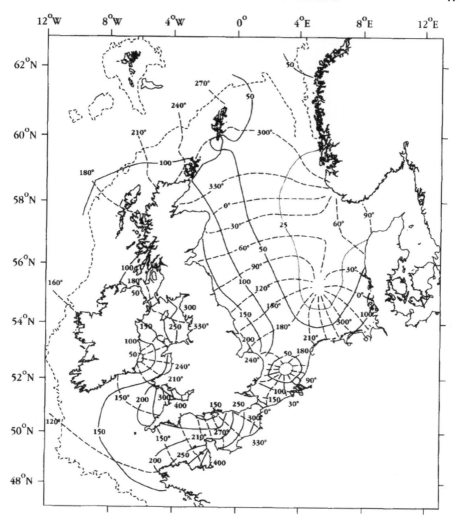

Figure 7.3 The amphidromic points: cotidal lines (broken lines in degrees) and corange lines (solid lines in cm) on the north-west European continental shelf (redrawn from Flather (1976) and Prandle (1997)).

It is worth noting that these locations correspond to the amphidromic points and thus exhibit classical standing wave flows.

The maximum amplitude of a mean Spring tidal current for the waters of the UK is mapped in (Figure 7.4) (DTI, 2008), and ABPMer's charts of the M_{U2} and S_{U2} semi-amplitudes are displayed in (Figure 7.5). These results are discussed further below.

Overall, the progress of the tidal wave is in a northerly direction from Gibraltar, up the west coasts of Spain, Portugal, and France, and then along the western shelf and around Scotland. Once in the North Sea, the tidal wave is responsible for the three amphidromic points discussed earlier and travels down the eastern coastlines of Scotland and England and then in a northerly direction along the coastlines

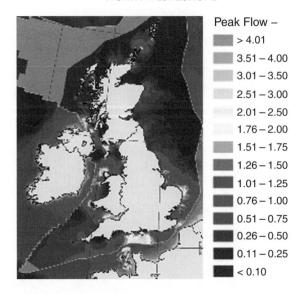

Figure 7.4 The peak mean Spring tide current speed ($m\,s^{-1}$) in UK waters (from DTI, 2008).

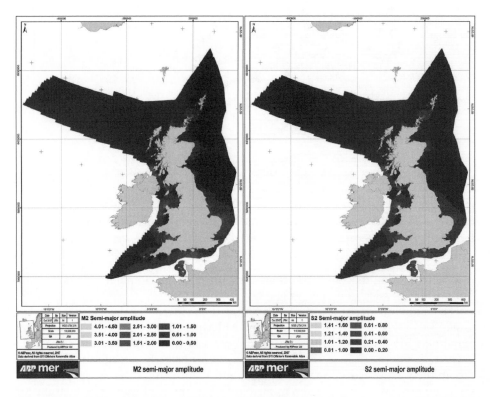

Figure 7.5 (Left) The M_{U2} semi-axis tidal-current amplitude and (right) the corresponding S_{U2} amplitudes. Colour copies of these images can be viewed at the book's website.

of the Low Countries and Norway. The movement of these elements of the tide is as a *progressive wave*, with peak flows corresponding to high and low water conditions. It is also generally true that the maximum current speed during the tidal cycle in progressive waves increases with the tidal range and hence with distance from the amphidromic point. Thus, a comparison of Figures 7.4 and 7.5 confirms that the maximum flows, for example, in the North Sea occur along the coastlines of England and Scotland, where the ranges increase with distance from the amphidromic point.

However, an entirely different process is operative in the Mediterranean, the English Channel, and the Irish Sea. Tides in the regions are dominated by a rocking motion, with high water occurring alternately at the open and effectively closed ends of the basin and a region in the centre where the tidal range is much smaller. These tidal responses are characteristic of a *standing wave* motion, and are confirmed by the existence of strong tidal streams towards the centres of the region (at the standing wave nodes), even though the tidal ranges are relatively small. Another characteristic of standing wave motion is that the maximum flows correspond to mid-water on the flood and the ebb, as opposed to the correspondence to high and low water in progressive waves. Estuaries also show the characteristics of standing wave motion, with peak currents corresponding to mid-tide.

7.4 Mediterranean

7.4.1 Introduction

The tides in the western Mediterranean are affected by the cooscillating tides entering through the Straits of Gibraltar. In contrast, the tides in the eastern Mediterranean are relatively independent of the ocean forcing. The interaction between the western and eastern sections of the Mediterranean is apparent. The largest tides in the Mediterranean are found in the Gulf of Gabes. The tides entering through the Strait of Gibraltar decay rapidly, and generally the tides are rather small in both basins of the Mediterranean. However, the tides are significant in the Aegean Sea. The tides in the Black Sea are rather small, principally because of its restricted connection to the Aegean through the narrow Straits of Bosphorous and Dardanelles and the Sea of Marmora. Consequently, the tides there are principally driven by direct astronomical forcing. Typically, the rotation of the Earth causes nodal lines to rotate cyclonically (counterclockwise sense) in the Northern Hemisphere, and in the opposite direction in the Southern Hemisphere. However, the anticyclonic amphidrome in the Black Sea results from forced tidal oscillations, even though the rotation of the Earth does act against it in a dissipative fashion.

7.4.2 Strait of Messina

The almost landlocked Mediterranean, disconnected as it is from the tidal engine of the north Atlantic, has tidal ranges measured in decimetres and minimal tidal currents, except perhaps for certain straits through which flows are locally accelerated. Thus, in the Strait of Messina, between Italy and Sicily, there are tidal currents that have been measured to exceed $9\,\mathrm{km\,h^{-1}}$ ($2.5\,\mathrm{m\,s^{-1}}$) (Massi, Salusti, & Stocchino,

1979), and therefore there exists the potential for tidal stream power developments. The Strait of Messina is a narrow channel separating the island of Sicily from the Italian peninsula, connecting the Ionian Sea with the Tyrrhenian Sea. In the Strait, strong tidal currents occur, which interact with the shallow sill located in its centre. The mean depth of the Strait is only 80 m, compared with depths of over 800 m to the south. However, internal wave trains of solitary waves can be generated, and these are evidenced as sea surface manifestations because they are associated with variable surface currents that modify the surface roughness. These can be detected as dark (smooth) and light (rough) patterns when the solar illumination is at the correct angle relative to the orientation of the satellite. These internal waves can significantly modulate the tidal-current regime. Nevertheless, attempts have already been made to extract tidal stream energy from the Strait, and these will no doubt continue in the future (Section 4.12).

7.4.3   The Bosphorous

On the very borders of Europe, in the narrow Bosphorous at Constantinople, which links the Mediterranean with the Black Sea, Özsoy, Latif, and Besiktepe (2002) report the results from current meter surveys showing that the normal two-layer Mediterranean directed flow on the surface and Black Sea directed flow at depth is modulated by semi-diurnal tidal forcing. Currents are shown to exceed $2.5 \, \text{m s}^{-1}$ at the surface, and to correlate with tidal forcing from the Mediterranean, but to be frequently 'blocked' by wind forcing from the Black Sea.

7.5   Spain and Portugal

There are very few tidal diamonds along the Spanish and Portuguese coasts (Figure 7.6), and, in general, the low current speeds (Fanjul, Gomez, & Sanchez-Arevalo, 1997; Sauvaget, David, & Guedes Soares, 2000) were confirmed by the numerical modelling described earlier (Marta-Almeida & Dubert, 2006). At certain river mouths, however, there are significant currents. For example, Station 7.1 at Rael de Santo Anotino exhibits a mean Spring tide peak tidal current of $1.3 \, \text{m s}^{-1}$, with which the 'standard' turbine in the STEM model generates $164 \, \text{MW h year}^{-1}$ at a rate of £0.919 MW h^{-1}. Further west, at Station 7.2 in the entrance to the Porto de Faro coastal lagoon, the currents peak at $1.1 \, \text{m s}^{-1}$ and would only generate $29 \, \text{MWh year}^{-1}$ in STEM, increasing the relative price to £5.113 MW h^{-1}. Finally, further north, Station 7.3 at Setubal (south of Lisbon) experiences peak currents of $1.2 \, \text{m s}^{-1}$, generating $1510 \, \text{MW h year}^{-1}$ at a relative price of £1.509 MW h^{-1}.

Such low outputs, even at the most promising sites, must be viewed against a background of variable state subsidy for renewable energy in general and tidal power in particular. At the present time (2007), Spain does not have any system of 'feed-in' tariffs (Cameron, 2007). Portugal, however, is encouraging marine power developments with a feed-in tariff for marine energy set at ECU 0.23 kW h^{-1} (2007 prices) and sympathetic planning and grid connection regulations.

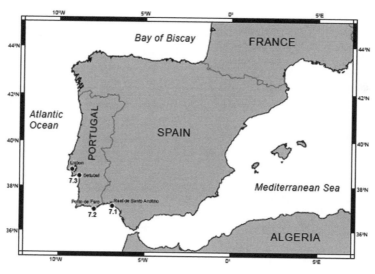

Figure 7.6 Spain and Portugal.

7.6 France

7.6.1 Introduction

Station 7.4 represents the offshore tidal currents in the southern part of the Bay of Biscay, peaking at little more than a few tens of centimetres per second. In the entrance to the bays and estuaries that form this coastline, however, stronger currents develop. For example, the coastal lagoon at Archachon (Station 7.5) exhibits much stronger currents peaking at $1.5\,\mathrm{m\,s^{-1}}$ on a mean Spring tide. Further north,

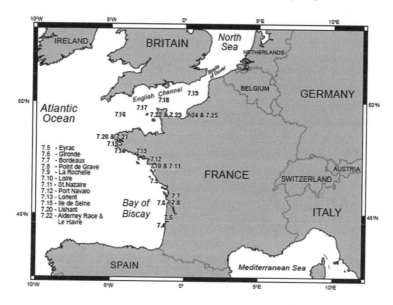

Figure 7.7 The Atlantic coastline of France.

at the entrance to the Gironde, downstream from Bordeaux, there are Stations 7.6 and 7.7 with peaks of about 2.1 m s^{-1}, while even inside the Point de Grave there are corresponding flows of 2.2 m s^{-1} at Station 7.8.

Further north, the currents are not so strong around La Rochelle, although the offshore coastal current increases to, for example, 0.4 m s^{-1} at Station 7.9. At the entrance to the Loire, however, off St Nazaire, where the mean Spring tidal range has increased to more than 5.0 m, the currents at Stations 7.10 and 7.11 are 1.7 and 1.1 m s^{-1} respectively on a mean Spring tide. A little further north, at the entrance to the coastal lagoon at Port Navalo, at Station 7.12, currents exceed 2.7 m s^{-1}, while at Lorient (Station 7.13) currents drop to less than 1.0 m s^{-1}.

Finally, for this section of the French coastline, the offshore currents increase to, for example, 0.8 m s^{-1} at Station 7.14 and accelerate to 2.7 m s^{-1} between the Ile de Sein and the headland at Station 7.15.

The French coastline along the southern shore of the English Channel shows a series of variations in tidal range and concomitant tidal-current speed. The offshore variations are, in turn, influenced by the local topography and regional bathymetric trends. The offshore current speeds show the patterns described earlier as the standing wave system in the English Channel is traversed, rising from, for example, less than 1.0 m s^{-1} at Stations 7.16 and 7.17, peaking at more than 2.0 m s^{-1} at Station 7.18, and reducing slightly as the Strait of Dover is approached at Station 7.19.

Closer inshore, the French coastline concentrates the flow around Ushant (Stations 7.20 and 7.21) to 1.8–1.9 m s^{-1}, Alderney Race (Stations 7.22 and 7.23) to about 3.0 m s^{-1}, and to a lesser extent in the entrance to the Seine at Le Havre (Stations 7.24 and 7.25), with peak mean Spring tidal currents of about 1.5 m s^{-1}. The main regions identified in this overview are described in more detail in the following sections.

7.6.2 The Gironde

The Gironde is formed from the meeting of the Dordogne and Garonne below the centre of Bordeaux and is approximately 65 km long and 3–11 km wide. The mean spring tidal currents at the mouth (Stations 7.6 and 7.7) peak at 2.1 m s^{-1}, and near Point de Grave there are corresponding flows of 2.2 m s^{-1} at Station 7.8. The currents were input into the standard STEM software and generated annual power outputs of 700–900 MWh year^{-1}, with a corresponding cost of about £0.16–0.18 kW h^{-1}.

7.6.3 Port Navalo

Station 7.12 at the entrance to the coastal lagoon of Port Navalo evidences a mean spring tide peak current that exceeds 2.7 m s^{-1}. The lagoon, through which the ancient township of Vannes was accessed from the Atlantic, is about 100 km^2 in surface area, and, together with the large tidal range of more than 4.0 m, generates a very considerable current because the entrance is less than 1 km wide. The 'standard' power generated with the STEM software at this site is more than 1800 MWh year^{-1}, and this corresponds to one of the lowest relative costings of £0.083 kW h^{-1}.

7.6.4 Ile de Sein

Between the Ile de Sein and the mainland, the offshore currents increase to 0.8 m s^{-1} at Station 7.14 and accelerate to a peak of 2.7 m s^{-1} at Station 7.15. The corresponding annual 'standard' power outputs also increase to more than 1600 MW h year^{-1}, so that the corresponding electricity generation costs in the STEM software are about £0.093 kW h^{-1}.

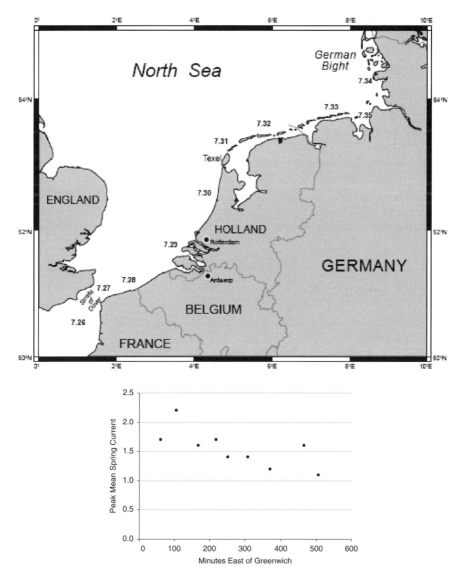

Figure 7.8 (Upper) The southern North Sea and (lower) the peak mean Spring tidal-current speed as a function of minutes east of Greenwich for Stations 7.26 to 7.34.

7.6.5   Alderney

The Alderney Race, where the Bay of Biscay meets the English Channel with a local tidal range of more than 6.0 m (Stations 7.22 and 7.23), concentrates the flow to more than 3.0 m s^{-1}. The STEM software suggests that this corresponds to more than 2200 MW h year^{-1} and a correspondingly low cost of £0.067 kW h^{-1}. Bahaj and Myers (2004) and Myers and Bahaj (2005) report simulations with a horizontal-axis marine current turbine using 2D aerofoil data, a blade element momentum software package, and tidal-current data from the Alderney Race. The optimum configuration predicted an annual energy output of 1340 GW h at a rated array capacity of approximately 1.5 GW.

7.7   Belgium, the Netherlands, and Germany

The transect of tidal-current speeds in a north-easterly direction along the coastlines of Belgium, the Netherlands, and Germany is represented by Stations 7.26 to 7.34. The locations are shown in Figure 7.8 (upper), and the data are summarized in Figure 7.8 (lower). It is apparent that the peak mean Spring tidal-current speeds generally decrease in an easterly direction from about 2 knots (1.0 m s^{-1}) to about 1 knot (0.5 m s^{-1}), except for the higher current speeds in the Channel and off the Elbe. None of these flows generates commercially viable outputs. Even though the entrance to the Elbe (Station 7.35) exhibits currents in excess of 1.4 m s^{-1}, the STEM software suggests an annual power output of less than 200 MW h year^{-1}, which is still not commercially viable.

7.8   Denmark and Norway

In the North Sea off the Norwegian coast, tidal currents are normally weak. For example, Figure 7.9 shows the locations of Stations 7.36 to 7.40, which evidence peak mean Spring currents of 0.4, 0.1, 0.1, 0.3, and 0.3 m s^{-1} respectively.

In coastal waters, however, tidal currents of the order of 1 m s^{-1} are common, with stronger currents evident at particular locations. The Maelstrom in the Lofoten Islands in northern Norway is one famous example. The Department of Mathematics, University of Oslo (1995), have a numerical model that shows the tidal current in the Drøbak Sound, a bottleneck of the Oslofjord, controlling the exchange of water masses between the inner and outer part of the fjord. Their second numerical model shows the tidal current in Trondheimsleia on the west coast of Norway. They also report similar models of the oscillating tidal current for other locations along the Norwegian coast and in the Barents Sea (Moe, Gjevik, & Ommundsen, 2000).

Gjevik (2004) and Gjevik et al. (2006) report further numerical modelling results for the tidal streams within the islands off the Norwegian coast.

In late 2002, Hammerfest Ström (see Chapter 4) deployed their horizontal-axis, three-bladed turbine in the Kvalsundet, a short distance from the city of Hammerfest (Figure 7.9). The location has a maximum current speed of 2.5 m s^{-1}, and the demonstrator is connected to the local grid (e.g. Hammerfest Ström, 2002).

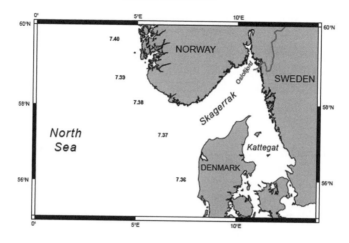

Figure 7.9 Denmark and Norway.

7.9 United Kingdom

7.9.1 Introduction

Moore (2007a) reviews a number of published investigations into the potential for tidal stream power in UK waters:

- **UK Atlas of Offshore Renewable Energy Resources** was published by ABPMer (2004) on behalf of the Department of Trade and Industry (DTI) and aimed to provide an overview of the wind, wave, and tidal potential in UK waters. The tidal resource was derived from the Proudman Oceanographic Laboratory (POL) high-resolution continental-shelf (HRCS) numerical model at a resolution of 1 nautical mile (approximately 1.8 km), but did not take account of the constraints of deploying the devices. The Atlas identified 12 areas of concentrated tidal stream power potential, which are listed in Table 7.1.

Table 7.1 Main areas for tidal stream power (ABPMer, 2004)

1	Orkney Islands
2	Pentland Firth
3	Humber
4	Norfolk
5	Dover
6	Isle of Wight
7	Portland
8	Channel Islands
9	Severn Estuary
10	Anglesey
11	Isle of Man
12	North Channel

- **Phase II UK Tidal stream Energy Resource Assessment.** This study by Black & Veatch (2005) as part of the Marine Energy Challenge on behalf of the Carbon Trust followed the Phase I report (Black & Veatch, 2004) and introduced the flux method of calculating the tidal resource (cf. Chapter 4) and the concept of the significant impact factor (SIF) which limited the extraction to 20 % of the potential power. The study concluded that the total extractable UK resource is about 18 TW h year^{-1}.

- **Variability of UK Marine Resources** characterized 36 sites around UK waters and assumed a device cut-in velocity of 1 m s^{-1}, 45 % efficiency, and a rated velocity at 70 % of the mean Spring current peak (MSCP) speed. The project outputs are focused towards the temporal variability in generated energy peaks compared with electricity demand. The study suggested that 3 GW of wave and tidal power could be installed by 2020 (Environmental Change Institute, 2005).

- **Wales Marine Energy Site Selection** investigated the practicable offshore wind, wave, and tidal energy resources around the Welsh coast. The study utilized the DTI Atlas and considered constraints, which included grid connection, environmental designations, and the Ministry of Defence. The work concluded that there were two potential sites for shallow water (<30m LAT) development: at Pembrokeshire (1.4 km^2 and 110 MW) and in the Bristol Channel (10 km^2 and 800 MW). The study also highlighted three deepwater sites at Anglesey (176 km^2 and 14 080 MW), Pembrokeshire (0.6 km^2 and 40 MW) and in the Bristol Channel (8 km^2 and 640 MW) (Project Management Support Services, 2006).

- **The Path to Power.** This three-stage study (ABPmer, 2006a,b; Bond Pearce, 2006; Climate Change Capital, 2006) considered constraints on development and concluded that the long-term tidal stream power potential would be around 3–5 % of UK demand, and that financing, grid access, permitting, and planning were the main obstacles to development.

- **Potential Impacts of Wave and Tidal Energy Extraction by Marine Renewables Development** reviewed the potential for marine renewables in Welsh territorial waters and considered environmental impacts (ABPmer, 2005, 2006a).

- **Scottish Marine Renewable Strategic Environmental Assessment** concluded that Scotland could achieve 10 % of its electricity demand from wave and tidal technologies, but that there were potential environmental impacts (Faber, Munsell, & Metoc, 2007).

- **UK Tidal Resource Review** considered the tidal stream power resource available from further analysis of the DTI Atlas, and, in the light of grid connection, policy issues and sea level rise (Metoc, 2007).

7.9.2 Moore's 2007 Analysis

The British sections of this chapter are based upon the excellent report recently produced by the English ABPMer consultancy for the British Department of Transport and Industry (Moore, 2007a, 2007b). The work is now in the public domain at www.renewables-atlas.info.

Moore (2007a) considered 35 potential tidal stream power devices (cf. a listing in Chapter 5), of which 21 had an appropriate level of information to be included. The devices were classified into five types:

- **type I** – horizontal axial-flow, single- or bidirectional turbines, fixed turbine direction;

- **type II** – horizontal axial-flow, multiple-direction turbines (yaw to accommodate flow reversals);

- **type III** – vertical-axis crossflow turbines that rotate regardless of flow direction;

- **type IV** – oscillating hydrofoil;

- **type V** – air injection technology.

However, it is arguable that type V, represented only by the Hydroventuri system (Chapter 4), may not be a true tidal stream power device, requiring, as it does, a pressure head such as induced by some form of barrier.

The physical resources of the United Kingdom marine areas were then considered, and Moore (2007a) showed that the overall area for deployment for type I and type III were similar, with technologies from both types having the potential for deployment in 3 % of UK waters. These types were found to have much larger physical area for deployment than type II (1 %), type IV (0.5 %), and type V (0.8 %). Moore concludes that, owing to overlap, some 28 000 km^2 of UK waters, representing about 3.1 % of the total, is potentially exploitable as a physical resource. However, his GIS-based analysis then continues to consider the various constraints. Twelve existing users or uses were considered as constraints and resulted in the loss of 12 % of the potential area. Marine cables account for over one-half of the constrained area, while aggregate winning and pipelines together exclude a half of the remaining amount. The constraints are listed in Table 7.2 and do not appear to include navigation.

It is not clear from the publication precisely how the potential power was calculated in each of the 1.8 km^2 cells considered. There is a description of the utilization of the mean Spring current speed 'derived by adding the M2 and S2 (*sic*) components of the tide in each data cell'. Elsewhere, the report claims that 'within the project database additional information is available regarding the precise tidal flows throughout an average year', without specifying the form of the additional information.

Moore (2007a) calculated the annual potential yield, with allowance for the Betz law coefficient, an average turbine efficiency of about 40 %, the physical and economic constraints, and the spacing of devices. The bottom 25 % of the water column was excluded to avoid the 'benthic boundary layer', and likewise the top 3.5 m was excluded because of wave action. This now appears insufficient because, for example, devices in Pentland or off Islay or in the Alderney Race would routinely be subject to wave action, beneath which very significant wave-induced currents extend

Table 7.2 Constraints listed by Moore (2007a)

1	Oil and gas infrastructure
2	Cables
3	Pipelines
4	Disposal sites
5	Wrecks
6	Protected wrecks
7	Obstructions
8	Aggregate winning
9	Wind farms
10	Anchorages
11	Maintenance dredging
12	Fish farms

Table 7.3 The six main areas for tidal power exploitation identified by Moore (2007a)

Area		Area (km²)	Number of devices	Annual output (GW h year⁻¹)
A	Pentland Skerries	5	20	162
	South Pentland	16	61	555
	North Pentland	5	17	181
	Westray	5	13	177
B	SW Islay	6	30	179
	West Islay	21	112	584
C	Anglesey	9	47	238
D	Ramsey Island	6	35	183
E	Isle of Wight	6	30	161
F	Aldernay Race	63	204	1,937

to considerable depths. No allowance was made for the transmission problems or for downtime. The result was an estimated output of 94 TW h year^{-1} from 200 000 devices occupying some 11 000 km^2.

More pragmatically, Moore (2007a) suggests that the installation of 569 devices in the 10 most suitable areas grouped into the six regions shown in Table 7.3 and arranged in 30 MW farms with an installed capacity of about 1500 MW would generate some 4.3 TW h year^{-1}.

Each of these areas is dealt with separately in the following sections.

7.9.3 Pentland and Westray

Moore's (2007a, 2007b) analysis indicates that there are three areas in the Pentland Firth (Figure 7.10) and an area near Westray that have potential annual yields of 70–80 GW h year^{-1}. The region is also examined theoretically by Bryden and Couch (2005). Stations 7.41 to 7.43 form a line across the western entrance to the Pentland Firth and have slow current speeds. Further into the Firth, Stations 7.44 and 7.45, however, exhibit annual outputs with the STEM software of 265 and 1659 MW h year^{-1}. To the east of the Firth, Station 7.46 shows reduced values, and

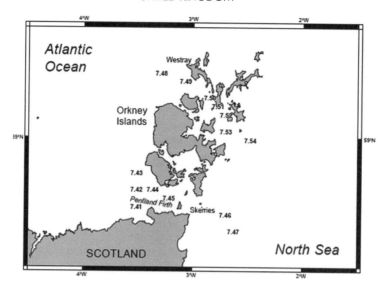

Figure 7.10 The Pentland Firth and Westray.

there appears to be a return to open water conditions with Station 7.47. The section through Westray given by Stations 7.48 to 7.54 shows a similar pattern, peaking at 2935 MW h year^{-1} in the channel and then reducing to open water values.

7.9.4 Islay

Moore's (2007a, 2007b) analyses indicate that there are two areas off the island of Islay that have potential annual yields of 70–90 GW h year^{-1} cell^{-1} and an overall potential annual output approaching 750 GW h year^{-1}. Admiralty (2007) has tidal diamonds in the region (Figure 7.11), for which data and analyses are given below. Stations 7.55 to 7.60 form a transect from north-west to south-east along the North

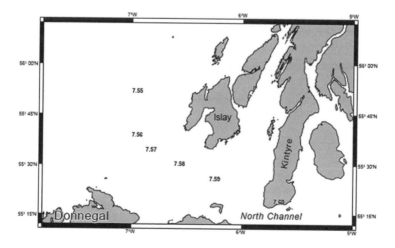

Figure 7.11 Islay.

Channel and are generally to the south and west of the cells identified by Moore (2007a). Nevertheless, there is a distinct pattern in the results. The maximum mean Spring currents are 1.3, 1.7, 1.3, 1.3, 1.4, and 1.7 m s^{-1} respectively. The transect from west to east shows outputs in STEM rising from about 120 MW h year^{-1} to a maximum of more than 450 MW h year^{-1}.

7.9.5   Anglesey

Moore's (2007a, 2007b) analyses indicate that there are two areas off the island of Anglesey with three cells that have potential annual yields of 70–90 GW h year^{-1} cell^{-1} and an overall potential annual output approaching 238 GW h year^{-1} from 47 devices. Admiralty (2007) has tidal diamonds in the region (Figure 7.12). Stations 7.61 to 7.64 form a loop around north-western Anglesey and lie generally through the cells identified by Moore (2007a). There is a distinct pattern in the results. The transect from south-west to north-east shows maximum mean Spring tidal currents of 1.1, 2.2, 3.1, and 1.7 m s^{-1} respectively. The corresponding annual power outputs in STEM rise from about 50 MW h year^{-1} to a maximum of 2378 MW h year^{-1}. Station 7.65 is in the entrance to the Menai Straits between Anglesey and the mainland and exhibits an annual power of 1382 MW h year^{-1}, with a corresponding current of 2.6 m s^{-1}.

7.9.6   Ramsey Island

Moore's (2007a, 2007b) analyses indicate that there are two cells off Ramsey Island that have potential annual yields of 70–90 GW h year^{-1} cell^{-1} and an overall potential annual output approaching 200 GW h year^{-1} from 35 devices. Admiralty (2007) has tidal diamonds in the region (Figure 7.13), for which data and analyses are shown below. Stations 7.66 to 7.68 form a loop around the south-western side of St David's Head and Ramsey Island and lie generally outside the cells identified by Moore (2007a). The transect from south to north shows low annual power outputs

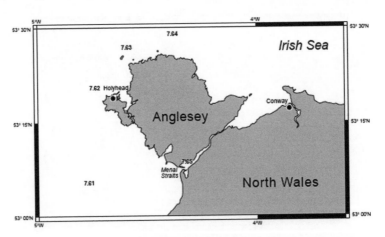

Figure 7.12   Anglesey.

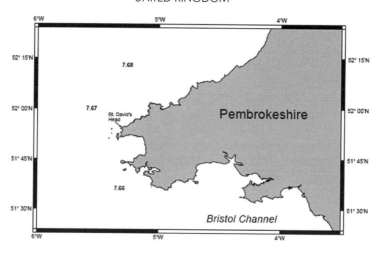

Figure 7.13 Ramsey Island.

in STEM of around 100–200 MW h year^{-1}, while the M_{U2} exhibits values of about 1.0 m s^{-1}.

7.9.7 Isle of Wight

Moore's (2007a, 2007b) analyses indicate that there are two cells off the Isle of Wight that have potential annual yields of 70–90 GW h year^{-1} cell^{-1} and an overall potential annual output approaching 161 GW h year^{-1}. Admiralty (2007) has tidal diamonds in the region (Figure 7.14). Stations 7.69 to 7.74 transect from west to east off the coast of the Isle of Wight and lie generally through the cells identified by Moore (2007a). There is a distinct pattern in the results. The transect from

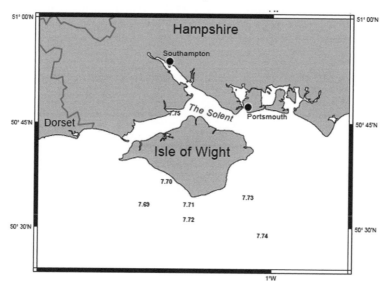

Figure 7.14 The Isle of Wight.

west to east shows annual power outputs in STEM rising to a maximum of more than 1000 MW h year^{-1} at Station 7.72, while the maximum mean Spring current varies between about 1.1 m s^{-1} and 2.3 m s^{-1}. Station 7.75 lies within the Solent between the Isle of Wight and the mainland and exhibits an annual power of about 415 MW h year^{-1}, with an M_{U2} of 1.35 m s^{-1}.

7.9.8 Alderney Race

Moore's (2007a, 2007b) analyses indicate that there are 20 cells in the UK waters within the Alderney Race that have potential annual yields of 70–90 GW h year^{-1} cell^{-1} and an overall potential annual output approaching 161 GW h year^{-1} from 30 devices. Admiralty (2007) has tidal diamonds in the region (Figure 7.15), for which data and analyses are shown below.

Stations 7.76 to 7.81 lie across the transect through the Race (from west to east), as indicated by the solid line in Figure 7.19. The transect from west to east shows annual power outputs in STEM of 2096, 1776, 3756, 3359, 5140, and 1176 MW h year^{-1} respectively. The maximum mean spring current varies between about 2.4 m s^{-1} and 4.9 m s^{-1}.

7.9.9 Ireland

The tidal wave moves as a progressive wave up the west coast of Ireland, and the currents that are generated are relatively small, except for local cases where sea lochs or bays have a narrow entrance. In general, Admiralty (2007) shows a small number of tidal diamonds to the west of this region (Figure 7.16), and a larger number in the Irish Sea to the east. Analyses of these data are discussed below. Three devices were considered by Bryan *et al.* (2004) and Bryan *et al.* (2006) in their analysis of the potential for tidal stream power in this region. They estimated that an average output of 130 MW might be generated at a few suitable locations using the commercial devices. Following advances in device size, they argued that it may be possible to achieve an average output of 325 MW

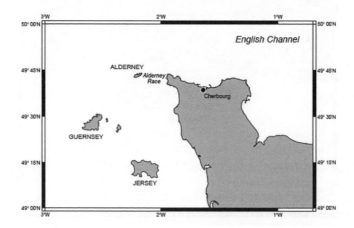

Figure 7.15 Alderney Race.

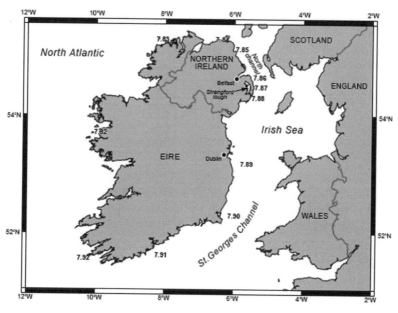

Figure 7.16 Ireland.

over a greater area. This would lead to a greater ability to reduce diurnal variation in supply by spreading the feasible tidal generation sites over differing tidal phases.

Stations 7.82 to 7.88 encircle Northern Ireland from west to east and show a pattern of change. For example, the maximum mean Spring currents are 0.2, 0.4, 1.3, 2.3, 1.8, 1.3, and 0.7 m s^{-1}. The corresponding annual power outputs in STEM rise to a maximum 1283 MW h year^{-1} at Station 7.85 but decrease both west and east from this location. Stations 7.89 to 7.92 encircle Ireland from north-east to south-west and show a reduction in currents and potential tidal power. The maximum mean Spring currents are 1.8, 1.4, 0.5, and 0.40 m s^{-1}. The corresponding annual power outputs in STEM reduce to the south and west from 481 MW h year^{-1} at Station 7.89 effectively to zero at Station 7.92.

Marine Current Turbines (cf. Chapter 4), however, installed their twin-turbine device in Strangford Lough during 2008, where the Spring tidal-current speeds can achieve very high rates in the narrow entrance.

7.10 The estuaries

7.10.1 The Humber

The Humber Estuary on the east coast of England is shown in Figure 7.17a. The annual standard power outputs generated in the STEM software for Stations 7.93 and 7.94 are 1800 and 1030 MW h year^{-1}, and the corresponding electricity costs are £0.084 and £0.146 kW h^{-1}.

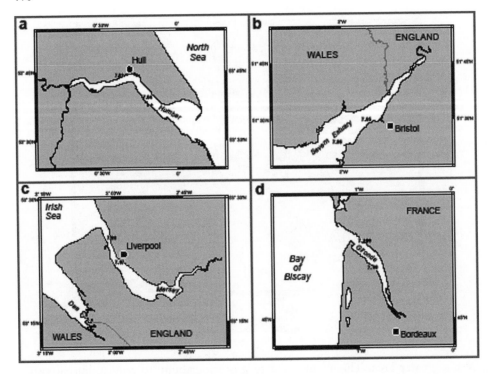

Figure 7.17 The estuaries: (a) the Humber; (b) the Severn; (c) the Mersey; (d) the Gironde.

7.10.2   The Severn

The Severn Estuary at the head of the Bristol Channel is shown in Figure 7.17b. The annual standard power outputs generated in the STEM software for Stations 7.95 and 7.96 are 1034 and 558 MW h year^{-1}, and the corresponding electricity costs are £0.145 and £0.269 kW h^{-1}.

7.10.3   The Mersey

The Mersey Estuary in north-west England is shown in Figure 7.17c. The annual standard power outputs generated in the STEM software for Stations 7.97 and 7.98 are 1119 and 1651 MW h year^{-1} respectively, and the corresponding electricity costs are £0.134 and £0.091 kW h^{-1} rspectively.

7.10.4   The Gironde

The Gironde on the north-east coast of France is shown in Figure 7.17d. The annual standard power outputs generated in the STEM software for Stations 7.99 and 7.100 are 219 and 595 MW h year^{-1} respectively, and the corresponding electricity costs are £0.687 and £0.253 kW h^{-1} respectively.

7.11 Bibliography

ABPMer (2004) *DTI Atlas of the Offshore Renewable Energy Resource*, DTI.

ABPMer (2005) *Potential Nature Conservation and Landscape Impacts of Marine Renewable Energy Development in Welsh Territorial Waters*. Countryside Commission for Wales.

ABPMer (2006a) *The Potential Nature Conservation Impacts of Wave and Tidal Energy Extraction by Marine Renewables Developments*. Countryside Commission for Wales.

ABPmer (2006b) *Path to Power Stage II: The Stakeholder/Statutory Bodies' View on Development*. BWEA npower Juice.

Bahaj, A.S. & Myers, L. (2004) Analytical estimates of the energy yield potential from the Alderney Race (Channel Islands) using marine current energy converters. *Renewable Energy* 29, 1931–1945.

Belderson, R.H. & Stride, A.H. (1966) Tidal current fashioning of a basal bed. *Mar. Geol.* 4, 237–257.

Bishop, P. & Jones, E.J.W. (1979) Patterns of glacial and post-glacial sedimentation in the Minches, North-West Scotland, in *The North West European Shelf Seas: The Sea Bed and Sea in Motion, Vol. I*, ed. by Banner, F.T., Collins, M.B., & Massie, K.S. Elsevier, Amsterdam, The Netherlands, pp. 89–104.

Black & Veatch (2004) *Tidal Stream Energy Resource and Technology Summary. Marine Energy Challenge*. The Carbon Trust. http://www.carbontrust.co.uk/NR/rdonlyres/CF053293-72CB-4204-B4DA-418AFD1244E2/0/PhaseIITidalStreamResourceReport.pdf [accessed 27 July 2007].

Black & Veatch (2005) *Phase II UK Tidal Stream Energy Resource Assessment Marine Energy Challenge*. The Carbon Trust. http://www.carbontrust.co.uk/NR/rdonlyres/19E09EBC-5A44-4032-80BB-C6AFDAD4DC73/0/TidalStreamResourceandTechnologySummaryReport.pdf [accessed 27 July 2007].

Bond Pearce (2006) *Path to Power Stage I: Wave and Tidal Energy Around the UK Legal and Regulatory Requirements*. BWEA npower Juice.

Bowden, K.F. (1979) Physical and dynamical oceanography of the Irish Sea, in *The North West European Shelf Seas: The Sea Bed and Sea in Motion, Vol. I*, ed. by Banner, F.T., Collins, M.B., & Massie, K.S. Elsevier, Amsterdam, The Netherlands, pp. 89–104.

Bryan, A.G., Fox, B., Crossley, P., & O'Malley, M. (2006) *Impact of Tidal Generation on Power System Operation in Ireland*. Power Engineering Society. http://ieeexplore.ieee.org/xpl/freeabs_all.jsp?arnumber=1708989 [accessed 10 August 2007].

Bryan, A.G., Fox, B., Crossley, P., & Whittaker, T.J.T. (2004) Tidal energy resource assessment for the Irish grid. *Universities Power Engineering Conference*, pp. 614–617.

Bryden, I.G. & Couch, S.J. (2005) ME1: marine energy extraction: tidal resource and analysis. *Renewable Energy* 31, 131–139, via http://www.sciencedirect.com/science/journal/09601481 accessed 8 July 2008.

Cameron, A. (2007) Marine means business. *Renewable Energy World* 10, 160–167.

Climate Change Capital (2006) *The Path to Power*. BWEA npower Juice.

Cooper, L.H.N. (1948) A submerged ancient cliff near Plymouth. *Nature* 161, 280.

Department of Mathematics, University of Oslo (1995) http://www.math.uio.no/~bjorng/tidevanns-modeller/tidemod.html [accessed 27 July 2007].

Doodson, A.T., and R.H.Corkn, 1932. New Tidal Charts for British Waters. *Geog. Jnl.*, 79, 321–323.

DTI (2008) http://www.dti.gov.uk/energy/sources/renewables/renewables-explained/wind-energy/age 27403.html [accessed 21 August 2008].

Egbert, G., Bennett, A., & Foreman, M. (1994) TOPEX/Poseidon tides estimated using a global inverse model. *J. Geophys. Res.* 99(24), 821–852.

Environmental Change Institute (2005) *Variability of UK Marine Resource*. The Carbon Trust.

Evans, D., Whittington, R.J., & Dobson, M.R. (1986) *Tiree Sheet. 56° N 08° W* British Geological Survey. 1:250 000 Series Solid Geology.

Faber, Munsell, & Metoc (2007) Scottish marine renewable strategic environmental assessment environmental report. Scottish Executive.

Fanjul, E.A., Gomez, B.P., & Sanchez-Arevalo, I.R. (1997) A description of the tides in the Eastern North Atlantic. *Progress in Oceanography* 40, 217–244.

Flather, R.A. (1976) A tidal model of the north-west European continental shelf. *Memoires Societe Royale des Sciences de Liege* 6, 141–164.

Flinn, D. (1967) Ice front in the North Sea. *Nature* 215, 1151–1154.

Goddard, A. (1965) *Recherches de Geomorphologie en Ecosse du Nord-Ouest*. Société d'Editions de belles Lettres, Paris, France, 701 pp.

Gjevik, B., Hareide, D.,Lynge, B., *et al.* (2006) Implementation of high resolution tidal current fields in electronic navigational chart systems. *Marine Geodesy* **29**, 1–17.

Gjevik, B., Hareide, D., Lynge, B.K., *et al.* (2004) http://www.math.uio.no/eprint/appl_math/2004/03-04.pdf accessed 27 July 2007.

Hamilton, D. (1979) The geology of the English Channel, South Celtic Sea and Continental Margin, South-Western Approaches, in *The North West European Shelf Seas: The Sea Bed and Sea in Motion, Vol. I*, ed. by Banner, F.T., Collins, M.B., & Massie, K.S. Elsevier, Amsterdam, The Netherlands, pp. 61–87.

Hammerfest Strøm (2002) *The Tidal Power Plant at Kvalsundet*. http://www.e-tidevannsenergi.com/Press15Sept02.pdf [accessed 26 May 2008].

Hardisty, J. (1990) *The British Sea: an Introduction to the Oceanography and Resources of the North-west European Continental Shelf*. Routledge, London, UK, 272 pp.

Hardisty, J. (2008a) Power intermittency, redundancy and tidal phasing around the United Kingdom. *Geog. J.* **174**, 76–84.

Hátún, H., Sandø, A.B., Drange, H., *et al.* (2005) Influence of the Atlantic Subpolar Gyre on the Thermohaline Circulation. *Science* **309**, 1841–1844.

Huthnance, J.M. (1982) On one mechanism forming linear sand banks. *Est. Coast. Shelf Sci.* **14**, 79–99.

Lennon G.W. (1961) The deviation of the vertical at Bidston in response to the attraction of ocean tides. *Geophys. J. R. Astr. Soc.*, **6**,(1) 64–84.

Marta-Almeida, M. & Dubert, J. (2006) The structure of tides in the Western Iberian region. *Con. Shelf Res.* **26**, 385–400.

Massi, M., Salusti, E., & Stocchino, C. (1979) On the currents in the Strait of Messina. *Il Nuovo Cimento C.2*, pp. 543–548. http://www.springerlink.com/content/r82w347438h0/ [accessed 25 July 2007].

Metoc (2007) *UK Tidal Resource Review*. Sustainable Development Commission.

Moe, H., Gjevik, B., & Ommundsen, A. (2000) A igh resolution tidal model for the coast of Møre and Trøndelag, western Norway. Preprint Series, Department of Mathematics, University of Oslo.

Moore, J. (2007a) Quantification of exploitable tidal energy resources in UK waters. Report R10439, ABPMer, Southampton, UK. Text http://www.abpmer.co.uk/allnews1623.asp [accessed 22 July 2007].

Moore, J. (2007b) Quantification of exploitable tidal energy resources in UK waters. Report R10439, ABPMer, Southampton, UK. GIS figures http://www.abpmer.co.uk/allnews1623.asp [accessed 22 July 2007].

Myers, L.E. & Bahaj, A.S. (2005) Simulated electrical power potential harnessed by marine current turbine arrays in the Alderney Race. *Renewable Energy* **30**, 1713–1731.

Özsoy, E., Latif, M.A., & Besiktepe, Ş. (2002) The current system of the Bosphorous Strait based on recent measurements, in Proceedings of 2nd Meeting of Physical Oceanography of Sea Straits, Villefranche, France. http://www.ims.metu.edu.tr/cv/ozweb/ozsoy-vf.pdf [accessed 25 July 2007].

Pantin, H.M. & Evans, C.D.R. (1984) The Quaternary history of the central and southwestern Celtic Sea. *Mar. Geol.* **57**, 259–293.

Pinet, P.R. (1996) *Invitation to Oceanography*, 3rd edition. West Publishing Co., St Paul, MN.

Prandle, D. (1997) Tidal currents in shelf seas – their nature and impacts. *Prog. Ocean.* **40**, 245–261.

Project Management Support Services (2006) *Wales Marine Energy Site Selection*. Welsh Development Agency.

Proudman, J. & Doodson, A.T. (1924) The principal constituent of the tides of the North Sea. *Phil. Trans. R. Soc. Lond. A* **224**, 185–219.

Puillat, I., Lazure, P., Jégou, A.M., *et al.* (2004) Hydrographical variability on the French continental shelf in the Bay of Biscay, during the 1990s. *Continental Shelf Research* **24**(10), 1143–1163.

Sauvaget, P., David, E., & Guedes Soares, C. (2000) *Coastal Engineering* **40**, 393–409.

Ting, S. (1937) The coastal configuration of western Scotland. *Geog. Annal.* 62–83.

Wood, A. (1976) Successive regressions and transgressions in the Neogene. *Mar. Geol.* **22**, M23–M29.

7.12 Appendix: STEM outputs for north-west Europe

Table 7A.1

Ref	Adm. Ref	Lat	Long	Pk Sp kt	Pk Np kt	Dist km	F_{zu}	M_2 ms⁻¹	S_2	K_2	Mean P kW	Max P kW	Output MWh/y	Cap Fact	Cost £m	£/kWh
7.1	SN175B	37 11.0N	07 24.4W	2.6	1.8	0.0	0	1.12	0.20	0.22	19	141	164	0.02	15,000	0.919
7.2	SN175D	36 58.7N	07 52.1W	2.1	1.0	0.0	0	0.79	0.56	0.16	12	131	107	0.01	15,000	1.407
7.3	SN174A	38 29.5N	08 55.7W	2.4	1.6	0.0	0	1.02	0.20	0.20	11	110	100	0.01	15,000	1.509
7.4	SN169E	44 31.3N	01 18.6W	0.2	0.1	0.0	0	0.08	0.03	0.02	0	0	0	0.00	15,000	inf
7.5	SN169A	44 40.3N	01 10.2W	3.0	1.5	0.0	0	1.15	0.38	0.23	26	207	227	0.03	15,000	0.664
7.6	SN168P	45 39.5N	01 12.3W	4.2	2.4	0.0	0	1.68	0.46	0.34	93	578	819	0.09	15,000	0.184
7.7	SN168AL	45 39.1N	01 09.7W	4.1	2.4	0.0	0	1.66	0.43	0.33	88	539	771	0.09	15,000	0.195
7.8	SN168V	45 32.8N	01 02.1W	4.4	2.3	0.0	0	1.71	0.54	0.34	103	656	900	0.10	15,000	0.167
7.9	SN167CL	46 30.9N	01 59.1W	0.8	0.4	0.0	0	0.31	0.10	0.06	0	0	0	0.00	15,000	inf
7.10	SN167W	47 15.9N	02 11.9W	3.3	2.3	0.0	0	1.43	0.26	0.29	49	289	433	0.05	15,000	0.347
7.11	SN167AY	47 16.4N	02 10.4W	2.2	1.4	0.0	0	0.92	0.20	0.18	6	84	52	0.01	15,000	2.864
7.12	SN167BV	47 33.7N	02 54.9W	5.2	3.5	0.0	0	2.22	0.43	0.44	206	1000	1809	0.21	15,000	0.083
7.13	SN165K	47 44.2N	03 20.9W	1.8	1.0	0.0	0	0.71	0.20	0.14	0	45	4	0.00	15,000	36.827
7.14	SN165AP	47 50.0N	04 30.0W	1.2	0.8	0.0	0	0.51	0.10	0.10	0	0	0	0.00	15,000	inf
7.15	SN164BT	48 02.5N	04 46.7W	5.2	3.0	0.0	0	2.09	0.56	0.42	184	1000	1611	0.18	15,000	0.093
7.16	SN163B	49 15.7N	04 44.5W	1.9	1.0	0.0	0	0.74	0.23	0.15	1	53	10	0.00	15,000	14.788
7.17	SN002M	49 37.1N	03 13.7W	1.6	0.8	0.0	0	0.61	0.20	0.12	0	0	0	0.00	15,000	inf
7.18	SN160F	49 59.2N	01 37.1W	4.2	2.1	0.0	0	1.61	0.54	0.32	86	567	756	0.09	15,000	0.199
7.19	SN158J	50 13.6N	00 30.1E	2.3	1.3	0.0	0	0.92	0.26	0.18	7	95	64	0.01	15,000	2.365
7.20	SN163J	48 31.4N	04 49.7W	3.6	1.8	0.0	0	1.38	0.46	0.28	51	357	450	0.05	15,000	0.334
7.21	SN163M	48 36.3N	04 45.8W	3.8	1.9	0.0	0	1.45	0.48	0.29	62	420	542	0.06	15,000	0.277
7.22	SN161BI	49 44.6N	01 55.8W	6.3	4.3	0.0	0	2.70	0.51	0.54	342	1000	2999	0.34	15,000	0.050
7.23	SN161BM	49 44.4N	01 52.1W	5.9	3.3	0.0	0	2.35	0.66	0.47	255	1000	2239	0.26	15,000	0.067
7.24	SN159BA	49 26.1N	00 06.9E	3.3	1.4	0.0	0	1.20	0.48	0.24	35	270	303	0.03	15,000	0.496

Table 7A.1 (continued)

Ref	Adm. Ref	Lat	Long	Pk Sp kt	Pk Np kt	Dist km	F_{zu}	M_2 ms^{-1}	S_2	K_2	Mean P kW	Max P kW	Output MWh/y	Cap Fact	Cost £m	£/kWh
7.25	SN159BB	49 26.1N	00 08.7E	3.1	1.4	0.0	0	1.15	0.43	0.23	28	225	245	0.03	15,000	0.613
7.26	SN009D	50 48.5N	01 11.8E	1.7	0.8	0.0	0	0.64	0.23	0.13	0	0	0	0.00	15,000	inf
7.27	SN009BA	51 04.9N	01 46.7E	2.2	1.5	0.0	0	0.94	0.18	0.19	7	85	57	0.01	15,000	2.615
7.28	SN012AA	51 23.3N	02 47.9E	1.6	1.2	0.0	0	0.71	0.10	0.14	0	0	0	0.00	15,000	inf
7.29	SN156W	51 52.4N	03 38.7E	1.7	1.1	0.0	0	0.71	0.15	0.14	0	39	0	0.00	15,000	inf
7.30	SN150AB	52 35.0N	04 12.0E	1.4	1.0	0.0	0	0.61	0.10	0.12	0	0	0	0.00	15,000	inf
7.31	SN148G	53 30.0N	05 08.0E	1.4	1.1	0.0	0	0.64	0.08	0.13	0	0	0	0.00	15,000	inf
7.32	SN147T	53 40.0N	06 10.0E	1.2	0.7	0.0	0	0.48	0.13	0.10	0	0	0	0.00	15,000	inf
7.33	SN146S	53 50.6N	07 46.1E	1.6	1.1	0.0	0	0.69	0.13	0.14	0	0	0	0.00	15,000	inf
7.34	SN144Q	54 14.0N	08 26.9E	1.1	0.8	0.0	0	0.48	0.08	0.10	0	0	0	0.00	15,000	inf
7.35	SN145W	53 40.6N	08 04.3E	2.7	1.8	0.0	0	1.15	0.23	0.23	21	157	186	0.02	15,000	0.809
7.36	SN142AE	55 00.0N	07 00.4E	0.8	0.6	0.0	0	0.36	0.05	0.07	0	0	0	0.00	15,000	inf
7.37	SN128F	57 10.0N	05 53.9E	0.2	0.1	0.0	0	0.08	0.03	0.02	0	0	0	0.00	15,000	inf
7.38	SN128C	57 50.0N	04 46.0E	0.2	0.1	0.0	0	0.08	0.03	0.02	0	0	0	0.00	15,000	inf
7.39	SN127A	59 20.0N	03 34.0E	0.5	0.2	0.0	0	0.18	0.08	0.04	0	0	0	0.00	15,000	inf
7.40	SN079C	61 00.0N	02 32.0E	0.6	0.2	0.0	0	0.20	0.10	0.04	0	0	4	0.00	15,000	inf
7.41	SN0281	58 39.7N	03 29.2W	1.8	1.0	0.0	0	0.71	0.20	0.14	0	45	0	0.00	15,000	inf
7.42	SN028J	58 43.7N	03 28.3W	1.6	0.9	0.0	0	0.64	0.18	0.13	0	0	0	0.00	15,000	inf
7.43	SN028K	58 47.8N	03 27.9W	1.3	0.7	0.0	0	0.51	0.15	0.10	0	0	0	0.00	15,000	inf
7.44	SN028D	58 43.6N	03 22.7W	3.2	1.3	0.0	0	1.15	0.48	0.23	30	245	265	0.03	15,000	0.567
7.45	SN0280	58 43.6N	03 14.2W	5.4	2.7	0.0	0	2.07	0.69	0.41	189	1000	1659	0.19	15,000	0.091
7.46	SN027I	58 38.5N	02 45.6W	1.4	0.6	0.0	0	0.51	0.20	0.10	0	0	0	0.00	15,000	inf
7.47	SN027AH	58 34.4N	02 39.8W	1.1	0.5	0.0	0	0.41	0.15	0.08	0	0	0	0.00	15,000	inf

7.48	SN028C	59 16.0N	03 17.5W	0.8	0.4	0.0	0	0.31	0.10	0.06	0	0	0	0.00	15,000	inf
7.49	SN028B	59 13.7N	03 03.1W	3.0	1.2	0.0	0	1.07	0.46	0.21	24	201	208	0.02	15,000	0.724
7.50	SN027P	59 11.1N	02 52.4W	5.1	2.1	0.0	0	1.84	0.77	0.37	146	992	1285	0.15	15,000	0.117
7.51	SN027T	59 08.1N	02 48.4W	7.2	2.8	0.0	0	2.55	1.12	0.51	334	1000	2935	0.34	15,000	0.051
7.52	SN027X	59 06.1N	02 45.1W	4.2	1.7	0.0	0	1.50	0.64	0.30	78	553	688	0.08	15,000	0.219
7.53	SN027AD	59 02.2N	02 44.9W	2.2	0.9	0.0	0	0.79	0.33	0.16	4	80	39	0.00	15,000	3.816
7.54	SN027AG	59 58.5N	02 32.1W	1.8	0.7	0.0	0	0.64	0.28	0.13	0	43	2	0.00	15,000	inf
7.55	SN038G	55 47.4N	06 59.9W	2.6	1.3	0.0	0	0.99	0.33	0.20	14	135	120	0.01	15,000	1.252
7.56	SN038C	55 33.8N	06 59.3W	3.3	1.7	0.0	0	1.28	0.41	0.26	38	276	335	0.04	15,000	0.449
7.57	SN065N	55 29.7N	06 51.4W	2.5	1.5	0.0	0	1.02	0.26	0.20	13	123	111	0.01	15,000	1.355
7.58	SN065O	55 24.4N	06 06.0W	2.6	1.5	0.0	0	1.05	0.28	0.21	15	137	132	0.02	15,000	1.138
7.59	SN065C	55 22.9N	06 06.0W	2.8	1.4	0.0	0	1.07	0.36	0.21	19	168	171	0.02	15,000	0.881
7.60	SN041D	55 15.1N	05 37.3W	3.4	2.1	0.0	0	1.40	0.33	0.28	49	310	428	0.05	15,000	0.351
7.61	SN048B	53 05.5N	04 44.6W	2.2	1.3	0.0	0	0.89	0.23	0.18	6	83	49	0.01	15,000	3.072
7.62	SN048N	53 19.5N	04 41.9W	4.5	2.2	0.0	0	1.71	0.59	0.34	106	696	932	0.11	15,000	0.161
7.63	SN048M	53 25.1N	04 34.9W	6.2	3.1	0.0	0	2.37	0.79	0.47	271	1000	2378	0.27	15,000	0.063
7.64	SN048J	53 26.7N	04 20.4W	3.3	1.7	0.0	0	1.28	0.41	0.26	38	276	335	0.04	15,000	0.449
7.65	SN048G	53 07.5N	04 19.8W	5.1	2.5	0.0	0	1.94	0.66	0.39	158	1000	1382	0.16	15,000	0.109
7.66	SN050A	51 36.5N	05 17.1W	2.2	1.0	0.0	0	0.82	0.31	0.16	5	81	41	0.00	15,000	3.686
7.67	SN060A	52 00.3N	05 36.6W	2.9	1.3	0.0	0	1.07	0.41	0.21	21	184	187	0.02	15,000	0.803
7.68	SN060I	52 15.0N	06 16.6W	2.4	1.2	0.0	0	0.92	0.31	0.18	9	106	78	0.01	15,000	1.921
7.69	SN005AH	50 33.9N	01 29.3W	2.7	1.3	0.0	0	1.02	0.36	0.20	16	150	141	0.02	15,000	1.069
7.70	SN005AG	50 35.9N	01 23.0W	2.2	1.1	0.0	0	0.84	0.28	0.17	5	82	43	0.00	15,000	3.522
7.71	SN005AF	50 33.5N	01 16.7W	3.7	1.9	0.0	0	1.43	0.46	0.29	57	389	502	0.06	15,000	0.299
7.72	SN005AM	50 30.4N	01 04.5W	4.6	2.3	0.0	0	1.76	0.59	0.35	115	745	1011	0.12	15,000	0.149
7.73	SN005AE	50 34.4N	01 00.1W	2.7	1.3	0.0	0	1.02	0.36	0.20	16	150	141	0.02	15,000	1.069
7.74	SN005AL	50 28.0N	01 22.0W	2.8	1.4	0.0	0	1.07	0.36	0.21	19	168	171	0.02	15,000	0.881
7.75	SN005C	50 45.4N		3.5	1.8	0.0	0	1.35	0.43	0.27	47	329	415	0.05	15,000	0.362

Table 7A.1 (continued)

Ref	Adm. Ref	Lat	Long	Pk Sp kt	Pk Np kt	Dist km	F_{zu}	M_2 ms^{-1}	S_2	K_2	Mean P kW	Max P kW	Output MWh/y	Cap Fact	Cost £m	£/kWh
7.76	SN161BY	49 43.9N	02 04.5W	5.4	3.8	0.0	0	2.35	0.41	0.47	239	1000	2096	0.24	15,000	0.072
7.77	SN161A	49 43.9N	02 03.6W	5.6	2.6	0.0	0	2.09	0.77	0.42	202	1000	1776	0.20	15,000	0.085
7.78	SN161AB	49 42.3N	02 01.8W	7.6	4.3	0.0	0	3.03	0.84	0.61	428	1000	3756	0.43	15,000	0.040
7.79	SN161AC	49 40.6N	01 58.5W	6.6	4.6	0.0	0	2.86	0.51	0.57	383	1000	3359	0.38	15,000	0.045
7.80	SN161AA	49 42.9N	01 58.9W	9.7	5.8	0.0	0	3.95	0.99	0.79	586	1000	5140	0.59	15,000	0.029
7.81	SN161BO	49 43.6N	01 57.4W	4.7	2.7	0.0	0	1.89	0.51	0.38	134	810	1176	0.13	15,000	0.128
7.82	SN070A	53 40.4N	10 03.0W	0.4	0.2	0.0	0	0.15	0.05	0.03	0	0	0	0.00	15,000	inf
7.83	SN068A	55 26.5M	08 09.1W	0.7	0.4	0.0	0	0.28	0.08	0.06	0	0	0	0.00	15,000	inf
7.84	SN065O	55 25.4N	06 28.3W	2.6	1.5	0.0	0	1.05	0.28	0.21	15	137	132	0.02	15,000	1.138
7.85	SN065B	55 12.3N	06 02.3W	4.6	3.1	0.0	0	1.96	0.38	0.39	142	779	1243	0.14	15,000	0.121
7.86	SN064G	54 41.7N	05 28.2W	3.5	2.4	0.0	0	1.50	0.28	0.30	59	344	521	0.06	15,000	0.289
7.87	SN046A	54 32.2N	05 25.5W	2.3	1.2	0.0	0	0.89	0.28	0.18	7	94	60	0.01	15,000	2.493
7.88	SN063B	54 24.4N	05 21.4W	1.4	0.7	0.0	0	0.54	0.18	0.11	0	0	0	0.00	15,000	inf
7.89	SN062A	54 03.6N	05 44.5W	3.6	2.0	0.0	0	1.43	0.41	0.29	55	362	481	0.05	15,000	0.313
7.90	SN060E	52 21.5N	06 10.5W	2.8	1.6	0.0	0	1.12	0.31	0.22	21	171	187	0.02	15,000	0.803
7.91	SN075G	51 28.6N	08 47.0W	1.0	0.4	0.0	0	0.36	0.15	0.07	0	0	0	0.00	15,000	inf
7.92	SN074F	51 26.3N	10 14.8W	0.7	0.3	0.0	0	0.26	0.10	0.05	0	0	0	0.00	15,000	inf
7.93	SN017A	53 43.9N	00 20.9W	5.0	3.8	0.0	0	2.24	0.31	0.45	205	1000	1800	0.21	15,000	0.084
7.94	SN017E	53 38.2N	00 10.7W	4.4	2.8	0.0	0	1.84	0.41	0.37	117	675	1030	0.12	15,000	0.146
7.95	SN053D	51 30.5N	02 43.6W	4.6	2.4	0.0	0	1.79	0.56	0.36	118	749	1034	0.12	15,000	0.145
7.96	SN052L	51 23.2N	03 05.0W	3.8	2.0	0.0	0	1.48	0.46	0.30	64	423	558	0.06	15,000	0.269
7.97	SN045N	53 22.1N	02 58.5W	4.7	2.5	0.0	0	1.84	0.56	0.37	128	801	1119	0.13	15,000	0.134
7.98	SN045L	53 25.5N	03 01.0W	5.3	2.9	0.0	0	2.09	0.61	0.42	188	1000	1651	0.19	15,000	0.091
7.99	SN168Z	45 18.1N	00 46.1W	3.0	1.4	0.0	0	1.12	0.41	0.22	25	205	219	0.02	15,000	0.687
7.100	SN168U	45 35.2N	01 02.6W	3.8	2.2	0.0	0	1.53	0.41	0.31	68	429	595	0.07	15,000	0.253

8

North America

8.1 Introduction

This chapter considers the North American waters from the east coasts of Canada and the United States to the Gulf of Mexico and the Caribbean, and then from Alaska and the west coast of Canada and the United States to the Pacific coastlines of Central America (Figure 8.1). As was the case in the preceding chapter, it is recognized that much of this area can be discounted for the present purposes. This chapter presents a generalized overview, and then concentrates on a relatively small number of sites in local, confined straits that have been identified by previous researchers.

The Analysis of Tidal Stream Power   Jack Hardisty
© 2009 John Wiley & Sons, Ltd

Figure 8.1   North America.

1. Knick Arm	5. Head Harbour Passage	9. Bay of Fundy
2. Tacoma Narrows	6. Western Passage	10. Chesapeake Bay
3. Golden Gate	7. Muskeget Channel	11. New York
4. Minas Passage	8. Gulf of Maine	12. Florida

The geography and oceanography of the regions are described. The standardized inputs to STEM that were developed in Chapter 5 and applied in Chapter 7 to north-west Europe are utilized, alongside some very recent research reported in the literature, to investigate the potential for, and relative cost of, tidal stream power developments in North America. The conclusion has to be that certain local sites represent great potential for tidal stream power developments, but, again, these may in the short term be hindered by the costs of the early-stage technology and the installation and grid connection problems. The STEM outputs are shown as an Appendix at the end of this chapter.

This chapter has been based, in part, upon the work of the Californian Electric Power Research Institute (EPRI) (Bedard, 2007). EPRI have produced a series of reports on the technical, regulatory, economic, and resource elements of tidal stream power in North America, which are listed in Table 8.1. This work resulted in the completion of a comparative table (shown here as Table 8.2), which is discussed in the appropriate sections below.

Additional material for the Canadian waters was derived from Cornett (2007), who concluded that there is the equivalent to 42 240 MW of power potentially available in Canadian waters, but that only a few per cent of the total may be practically recoverable owing, in part, to problems with winter ice.

There is also a rich source of current meter data available in the tide tables now published by International Marine but containing the data obtained by the National

Table 8.1 EPRI reports on tidal stream power (hyperlinks given in the Bibliography are also available through the book's website)

Tidal in-stream energy reports
Bedard (2005)
Bedard & Siddiqui (2006)
DTA (2006a)
Hagerman (2006a)
Hagerman (2006b)
Hagerman (2006c)
Hagerman (2006d)
Hagerman & Polagye (2006)
Polagye (2006)
Polagye & Previsik (2006a)
Polagye & Previsick (2006b)
Polagye & Bedard (2006)
Previsic (2006a)
Previsic (2006b)
Previsic (2006c)
Previsic (2006d)
Previsic (2006e)
Previsic & Bedard (2005)
Bedard *et al.* (2006)

Tidal in-stream energy briefings
Bedard (2006a)
Bedard (2006b)
DTA (2006b)
EPRI (2005)

Table 8.2 The depth-averaged power density for the sites analysed by EPRI (Bedard, 2006a,b)

	Knik Point, AK	Washington	Golden Gate, CA	Western Passage, ME	Maine	Head Harbour Passage, NB	Minas Passage, NS
Cross-section (m^2)	72 500	62 600	74 700	17 500	36 000	60 000	225 000
Power density ($kW\ m^{-2}$)	1.6	1.7	3.2	0.95	2.9	1.0	4.5
Extractable power (MW)	17.4	16	35.5	2.0	15.6	10.0	152

Oceanographic and Atmospheric Administration (NOAA) from the National Ocean Services (NOS) network. In the present work, data were extracted from International Marine (1999), which published tidal-current tables for the Atlantic coast of North America, and International Marine (2006) for the Pacific coast of North America.

Seven coastal sites are identified (Table 8.2), each of which appears to be capable of generating more than 1000 MW h year^{-1} from the 'standardized' turbine at relative (2008) costs of less than £0.200 kW h^{-1}. There are three sites on the west coast: Knik Point in Alaska, Tacoma Narrows in Washington, and in the entrance to San Francisco Bay in California. There are four sites on the east coast: two sites in the United States – Muskeget Channel in Massachusetts and Western Passage in Maine, and two sites in Canada – Head Harbour Passage in New Brunswick and Minas Passage in Nova Scotia.

8.2   Geography of North America

8.2.1   East coast

The east coasts of Canada and the United States stretch from the Grand Banks of Newfoundland in the north to the keys of Florida and the Gulf of Mexico in the south. The coastline may be divided into four sections consisting of the following Provinces and States (Figure 8.2):

- **Canadian north-east**
  The Provinces of Newfoundland (NL), Nova Scotia (NS), and New Brunswick (NB).

- **Gulf of Maine and Bay of Fundy**
  The States of Maine (ME), New Hampshire (NH), and Massachusetts (MA).

- **New York and Chesapeake Bay**
  Richmond (RI), Connecticut (CT), New York (NY), New Jersey (NJ), Delaware (DE), Maryland (MD), and Virginia (VA).

- **Florida and the Gulf of Mexico**
  North Carolina (NC), South Carolina (SC), Georgia (GA), Florida (FL), Mississippi (MS), Louisiana (LA), and Texas (TX).

8.2.1.1 Canadian north-east

The Canadian north-east consists of the shallow areas of the Flemish Cap and the Grand Banks and the island of Newfoundland. Between the island and the mainland of Nova Scotia is the Lawrentian Channel leading into the St Lawrence Seaway connecting to the Great Lakes. The nearshore is called the Scotia Shelf and the whole area is called the Labrador Sea.

8.2.1.2 Gulf of Maine and Bay of Fundy

The Gulf of Maine extends from the Canadian border in the north to Nantucket Island in the south and is separated from the open North Atlantic by Georges and Browns Banks. The North-east Channel and the Great South Channel are the two

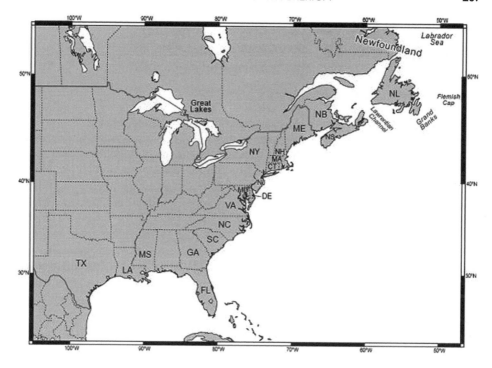

Figure 8.2 The eastern seaboard of North America and Canada.

main valleys that allow most of the water to exchange with the Gulf. Inside the Gulf, there are three deeper areas known as Wilkinson Basin, Georges Basin, and Jordan Basin. These also play a large part in the physical dynamics of the region (EET, 2007). The northern border of the Gulf of Maine merges into the Bay of Fundy which, at its upper end, separates into Chignecto Bay and the Minas Basin (see below) where the world's largest tidal ranges occur.

8.2.1.3 New York Bight and Chesapeake Bay

Between Nantucket Island in the north and Cape Hatteras in the south is the central North American continental shelf encompassing the New York Bight and the entrances to the Delaware Estuary and Chesapeake Bay. New York City, on the Hudson River, is at the western end of Long Island.

The New York Bight shelf is relatively wide, extending to some 180 km off New York. The Hudson Shelf Valley extends for the full width of the shelf in a northwest to south-east direction from the mouth of the Hudson River to the shelf edge, between Long Island and New Jersey (USGS, 2008).

South of New Jersey is the Delaware Estuary, which extends for some 200 km from its mouth at Browns Shoal to its head. South of the Delaware is Chesapeake Bay which is the largest estuary in the United States, exhibiting a surface area of approximately 11 400 km^2 and stretching 332 km from Virginia Beach to Havre de Grace at the mouth of the Susquehanna River. The Bay drains a region of some 165 800 km^2 (USGS 2007).

8.2.1.4 Florida and Gulf of Mexico

The Atlantic coast of the State of Florida extends from Cape Hatteras in the north to the Keys in the south. The northern area of this shelf is relatively wide but narrows in a southerly direction, and offshore is the Blake Plateau.

The Gulf of Mexico is an epicontinental basin largely surrounded by the North American continent and the island of Cuba. It is bounded on the north-east, north, and north-west by the Gulf Coast of the United States, on the south-west and south by Mexico, and on the south-east by Cuba. The shape of the basin is roughly oval, approximately 1500 km wide and is connected to the open ocean through the Florida Straits between the USA and Cuba, and with the Caribbean Sea via the Yucatan Channel between Mexico and Cuba. Tidal ranges are extremely small owing to the narrow connection with the ocean, and therefore the region is not considered further here.

8.2.2   West coast

The west coast of North America stretches from the ice-bound coastline of northern Alaska, through Canada, the western United States, and south to Mexico and Central America. The western seaboard of North America consists of four regions (Figure 8.3):

- **Alaska**
  The State of Alaska (AK).

- **Canadian north-west**
  The Provinces of the Yukon Territories (YK) and British Columbia (BC).

- **The west coast**
  Washington (WA), Oregon (OR) and California (CA).

- **West-central America**
  The countries of Mexico, Guatemala, El Salvador, Honduras, Nicaragua, Costa Rica, and Panama.

8.2.2.1 Alaska

The northernmost geographic area of Alaska is called the Arctic Coastal Plain. The coastal waters include Prudhoe Bay in the Arctic Ocean to the north and the Bering Sea between Hope Point and the Alaskan Peninsula and between Bering Islands and the Gulf of Alaska in the south. Although the Arctic coast shelf is relatively narrow,

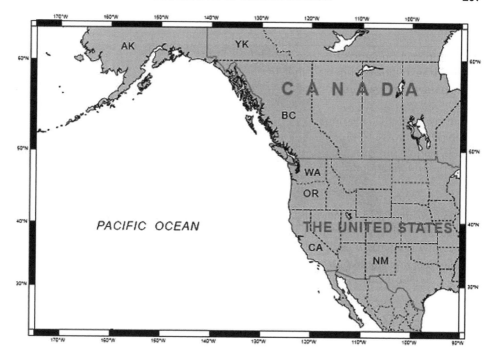

Figure 8.3 The western seaboard of North America and Canada.

the shelf edge roughly bisects the Bering Sea from north-west to south-east, with the shallow eastern and north-eastern waters constituting Bristol Bay. South of the Alaskan Peninsula, the shelf is once again narrow.

8.2.2.2 *Canadian north-west*

The Canadian north-west consists of the Provinces of the Yukon Territory and British Columbia and, offshore, has a narrow and featureless continental shelf. Although there are many islands off the Yucatan Territories, they are remote, and the possibilities for tidal stream power developments are, at best, some time away. Further south, however, the waters behind Vancouver Island on both the Canadian and United States side of the border have some very good potential sites, and these are discussed below.

Vancouver Island is the largest island on the west coast of North America. It stretches for 500 km south-east to north-west, with an area of 3 175 000 ha, and has 3460 km of coastline. The north-east shore of the Island is separated from the mainland by some narrow channels that open into the Strait of Georgia. It is separated from the Olympic Peninsula to the south by the Strait of Juan de Fuca, which itself connects to Puget Sound in Washington, as described below.

8.2.2.3 *The west coast*

The west coast consists of the States of Washington in the north, Oregon, and California. Puget Sound, now in Washington State, was explored by George Vancouver in 1792. The United States Geological Survey recognizes three entrances to the Sound: Admiralty Inlet, Deception Pass, and Swinomish Channel converging to the south on the Tacoma Narrows, a potential location for tidal stream power extraction, as detailed in Section 8.7.3 below. The Oregon coastline is bordered by Washington to the north and California to the south, and is particularly bereft of inlets and potential tidal stream power sites. The coastline of California is similar, but the great San Francisco Bay, with its strong tidal currents, has led to a number of stream power proposals, as detailed in Section 8.7.4 below.

The Pacific Ocean exhibits a geologically active fault boundary in this area, and the continental shelf is very narrow, often no more than 4 or 5 km wide. Along the California coast, the continental shelf and slope are etched by submarine canyons created by sporadic turbidity currents transporting detritus to the deep ocean (CERES, 2008).

8.2.2.4 *West-central America*

The Baja Peninsula in Mexico is 1200 km long, yet less than 100 km wide. Baja's Pacific coast is characterized by the upwelling of cool, deep ocean water. On the east, the Sea of Cortez reaches depths of more than 4000 m and, near the Colorado River Delta, exhibits tidal ranges of more than 8 m. Central America comprises seven countries: Belize, Guatemala, El Salvador, Honduras, Nicaragua, Costa Rica, and Panama, but these are not discussed further here.

8.3   Oceanography of North America

8.3.1   East coast

General circulation. The Gulf Stream system has three named components: the Florida Current, where it passes between Florida and the Bahamas, the Gulf Stream, where it flows along the coast of the USA, and the North Atlantic Current. The warm part of the Gulf Stream originates in the Gulf of Mexico, in which the circulation is first confined through the Yucatan Channel and then forms a loop before exiting into the Florida Current.

Salinity. Salinities are highest, at about 37 psu, in the central North Atlantic Gyre, and decrease towards the coast. There is also a slight decrease from about 36 psu in the south, off Florida, to about 34 psu towards the north and Canada.

Temperature. The Gulf Stream temperature (NCDC, 2008) is about 27 °C and decreases in a northerly and in an easterly direction along the Atlantic coast of North America and Canada.

Tidal range. Chapter 5 demonstrated that the tides on the east coast of North America are controlled by the Newfoundland amphidromic system in the north

Table 8.3 Spring tidal range on the Atlantic coast of North America

Name	Latitude	Spring range (m)
Cape Canaveral	28° 25′ N	1.2
Charleston	32° 47′ N	1.7
Hampton	36° 57′ N	0.8
Sandy Hook	40° 28′ N	1.6
Boston	42° 21′ N	3.1
St Johns	45° 16′ N	7.1

and the Bahamian amphidromic system in the south. Both tidal systems are rotating in an anticlockwise direction, so that the tidal wave (and the times of high water) progress in a southerly direction down the coastlines of north-east Canada and the United States, into the Gulf of Mexico. The tidal range decreases from more than 14.0 m in Cobequid Bay at the northern end of the Bay of Fundy as the amphidrome is approached towards Florida, and is less than 1.0 m in the partially landlocked Gulf of Mexico. Table 8.3 shows data for a number of sites taken from Admiralty (2007).

Tidal currents. The following is derived from International Marine (1999). The currents along the Atlantic coast of North America are semi-diurnal, with two floods and two ebbs in each day. In the Gulf of Mexico, however, the currents are daily in character. There are few Admiralty tidal diamond stations along this coast, so that reference is made to the local tide tables. Table 8.4 lists data entered into STEM for all of the east coast reference stations, using the mean Spring tide that occurred in the afternoon of 7 June 1999 and the mean Neap tide that occurred on the morning of 14 June 1999.

8.3.2 West coast

General circulation. The California Current, carrying water cooled by its passage through the northern latitudes, flows southward along the shore from the Washington – Oregon border to southern California. The current is modified by seasonal variations in wind direction that give three, more or less distinct, 'oceanic seasons'. Beginning in March, prevailing westerly winds, combined with the effects of the Earth's rotation, drive surface waters offshore. These waters are replaced by deep, cold water that flows up over the continental shelf to the surface, carrying with it dissolved nutrients from the decay of organic material that has sunk to the ocean floor. This process, known as upwelling, is restricted mainly to the west coasts of continents, and is responsible for the high productivity of California's nearshore waters.

The upwelling period continues until September, when north-westerly winds die down and the cold upwelling begins to sink. This period, characterized by relatively high surface temperatures, is known as the oceanic period, and generally lasts until October of each year.

Table 8.4 Estimates of maximum Spring and Neap tidal currents at the reference stations in International Marine (1999)

STEM		Spring (knots)	Neap (knots)
8.1	Aransas Pass, Texas	2.2	0.8
8.2	Baltimore Harbour Approach, Maryland	1.2	0.4
8.3	Bay of Fundy Entrance (Grand Manan Channel)	3.4	1.9
8.4	Bergen Point Reach, New York	2.3	1.6
8.5	Boston Harbour, Massachusetts	1.6	1.2
8.6	Cape Cod Canal, Massachusetts	4.8	3.7
8.7	Charleston Harbour, South Carolina	3.2	1.6
8.8	Chesapeake and Delaware Canal, Maryland	2.5	1.8
8.9	Chesapeake Bay Entrance, Virginia	1.8	0.6
8.10	Delaware Bay Entrance, New Jersey	1.9	1.2
8.11	Galveston Bay Entrance, Texas	3.8	0.8
8.12	East River, New York	5.1	3.2
8.13	Key West, Florida	2.5	0.7
8.14	Miami Harbour Entrance, Florida	2.2	1.4
8.15	Mobile Bay Entrance, Alabama	2.7	0.5
8.16	Old Tampa Bay Entrance, Florida	1.5	0.6
8.17	Pollock Rip Channel, Massachusetts	2.4	1.7
8.18	Portsmouth Harbour Entrance, New Hampshire	2.5	1.3
8.19	St Johns River Entrance, Florida	2.8	1.7
8.20	Savannah River Entrance, Georgia	2.7	1.8
8.21	Tampa Bay, Florida	2.1	0.5
8.22	Tampa Bay Entrance, Florida	2.6	0.6
8.23	The Narrows, New York Harbour, New York	2.6	1.6
8.24	The Race, Long Island Sound, New York	4.2	2.4
8.25	Throgs Neck, Long Island Sound, New York	1.2	0.6
8.26	Vieques Passage, Puerto Rico	1.1	0.5

In winter, changes in atmospheric conditions over the Pacific Ocean bring south-westerly winds to the California coast. In response to these winds, a northward surface current begins and flows along the coast inland of the California Current. This current, called the Davidson Current, generally lasts until February, when the prevailing winds shift again, and the cycle begins anew.

Salinity. Sea surface salinities of the west coast of North America are, in general, lower than at the corresponding latitudes, off the east coast. There are fresh areas with salinities of less than 34 psu at both the northern end, off Alaska, and the southern end, off the west American coastline, and coastal salinities only reach values of about 34 psu at maximum off southern California.

Temperature. The west coast of North America is, in general, cooler than the east coast waters at corresponding latitudes because of two factors. Firstly, the east coast is cooled by the colder waters of the south-flowing California Current, which represent the eastern side of the great, anticlockwise, North Pacific Gyre (see above). Secondly, and for similar reasons, the west coast does not evidence a warm, north-flowing current, such as the Gulf Stream, which is present in east coast waters. Thus, the 25 °C mean annual surface temperature only reaches as far north as Baja

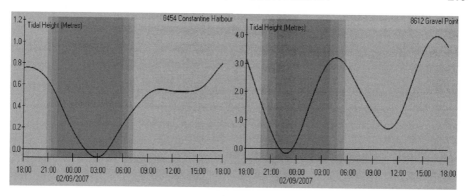

Figure 8.4 (Left) Tidal curve, Constantine Harbour, and (right) Gravel Point (Admiralty, 2007).

California on the west coast, while the corresponding contour is at much higher latitudes off Nantucket Island on the east coast.

Tidal range. The tides in the extreme north of the west coast are mixed, and in some areas diurnal. For example, at Constantine Harbour (51° 29′ N, 117° 01′ E) on the Aleutian Islands, the tide is clearly diurnal (Figure 8.4 (left)). Admiralty (2007) lists the following data:

HAT 1.2 MHHW 0.7 MLHW 0.5 MSL 0.43 MHLW 0.5 MLLW 0.0

In comparison, the tides at Gravel Point (60° 28′ N, 145° 58′ W) return to a more regular semi-diurnal nature (Figure 8.4 (right)), and Admiralty (2007) lists

HAT 4.4 MHHW 3.6 MLHW 3.1 MSL 1.98 MHLW 1.2 MLLW 0.0 m

The tidal range is taken here as (MHHW − MLLW) = 3.6.

Further south, tides on the west coast of North America and Canada are controlled by the north-eastern Pacific amphidromic system. The tidal system is rotating in an anticlockwise direction, so that the tidal wave (and the times of high water) progress in a southerly direction down the coastlines of north-west Canada, the United States, and Central America. The tidal ranges generally decrease in a southerly direction as the amphidrome is approached towards Mexico. Table 8.5 shows a number of sites taken from Admiralty (2007).

Tidal currents. The following is derived from International Marine (2006). As there are few Admiralty tidal diamond stations along this coast, reference is here made to the local tide tables, which list data for the following stations and have been coded into the STEM North America software as Stations 8.27 to 8.49 in Table 8.6. Daily current maxima rather than mean Spring and Neap peak currents are given in the tide tables (International Marine, 2006), and therefore the mean peak Spring tide data were taken from the afternoon of 12 June 2006, and the mean peak Neap tide data were taken from the morning of 5 June 2006.

Table 8.5 Variation in tidal range on the west coast of
North America (Admiralty, 2007)

Name	Latitude	Spring range (m)
Gravel Point	60° 28′ N	3.6
Cape Ommaney	56° 10′ N	3.0
Vancouver Island	48° 24′ N	1.8
San Francisco	37° 49′ N	1.6
Bahia	24° 38′ N	1.3
Rio Colorado	31° 46′ N	7.3

Table 8.6 Estimates of maximum Spring and Neap tidal currents at the
reference stations in International Marine (2006)

STEM		Spring (knots)	Neap (knots)
8.27	Active Pass, British Colombia	5.6	1.7
8.28	Admiralty Inlet, Washington	3.7	0.7
8.29	Suisan Bay, California	2.4	0.6
8.30	Burrard Inlet, British Colombia	5.0	1.2
8.31	Carquinez Strait, California	3.5	1.3
8.32	Deception Pass, Washington	7.1	4.4
8.33	Golden Gate Bridge, California	3.3	1.0
8.34	Grays Harbour Entrance, Washington	4.0	0.7
8.35	Humboldt Bay Entrance, California	1.5	0.9
8.36	Isanotski Strait, Alaska	4.2	1.8
8.37	Kvichak Bay, Alaska	3.1	2.1
8.38	North Inian Pass, Alaska	7.0	1.3
8.39	Race Rocks, British Colombia	3.2	1.9
8.40	Richmond, California	3.1	2.8
8.41	Rosario Strait, Washington	3.1	0.7
8.42	San Diego Bay Entrance, California	2.5	0.5
8.43	San Francisco Bay Entrance, California	5.4	1.7
8.44	San Juan Channel, Washington	4.5	1.1
8.45	Sergius Narrows, Alaska	7.1	2.6
8.46	Seymour Narrows, British Colombia	13.9	5.5
8.47	Snow Passage Narrows, Alaska	3.7	1.3
8.48	Puget Sound, Washington	5.0	1.4
8.49	Wrangell Narrows, Alaska	3.7	1.0

8.4 East coast of Canada

8.4.1 Minas Passage, Nova Scotia

The Bay of Fundy is a large inlet of the Atlantic Ocean, measuring some 270 km
long and 50–80 km wide, between New Brunswick and south-western Nova Scotia.
The Bay forms a narrow bend extending to the north-east of the wide Gulf of Maine
(Defant, 1961). Its width varies little until it is bifurcated into two canal-like arms by

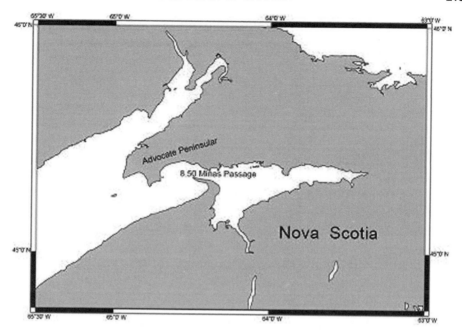

Figure 8.5 Minas Passage, Nova Scotia.

the Advocate Peninsula (Figure 8.5) which separates the Chignecto Basin to the west from the Minas Basin to the east. The water depths decrease from about 100m in the Bay to about 50m at the openings of the Basins. Consideration of the increasing tidal ranges and of the tidal lags demonstrates the presence of a standing wave which must be considered as a co-oscillating tide (Defant, 1961). At low tide, wide flats are laid bare, and the long estuaries of the rivers are drained.

Cornett (2007) shows a 'tidal-current power density' of 100 kW m^{-1} in the Minas Passage at the entrance to the Minas Basin, with mean Spring peak currents of about 4.5 m s^{-1} and corresponding mean Neap peak currents of about 2.5 m s^{-1}. These data are shown as Station 8.50 in Figure 8.8 and are entered into the North American STEM package. The results suggest that a 'standardized' STEM device generates power output of about 1000 MW h year^{-1} at a cost of about £0.15 kW h^{-1} for the Minas Passage.

Previsic (2006a) reports an alternative analysis of the potential for tidal stream power in Minas Passage, Nova Scotia, and Canada, in which the 'Nova Scotia electricity stakeholders' also 'selected' Minas Passage because 'it appeared to offer the best combination of tidal resource, grid interconnection, and access to engineering support'. Pervisic's (2006a) analysis uses a simpler synthesis of the tidal-current streams and estimates that the cross-sectional area of the site is 225 000 m^2, that the depth-averaged mean annual hydraulic power density (Chapter 3) is 4.5 kW m^{-2}, and that the resulting 'extractable' power (rated at 15 % of the total hydraulic power) is an annually averaged rate of 152 MW. This figure, however, appears to be the potential hydraulic power rather than the estimated power output, which would amount to some 45 %, or 70 MW, equivalent to about 600 GW h year^{-1}.

8.4.2   Head Harbour Passage, New Brunswick

Previsic (2006b) reports an analysis of the potential for tidal stream power in New Brunswick, Canada, for which the 'New Brunswick electricity stakeholders' selected Head Harbour Passage (Figure 8.6) because it appeared to offer the best combination of tidal resource, grid interconnection, and access to engineering support. Head Harbour Passage connects Passamaquoddy Bay in the Bay of Fundy to the open ocean.

Head Harbour Passage trends north-east to south-west from Friars Road and is the main shipping entrance channel to Passamaquoddy Bay. The north-west shore of Campobello Island and the entrance to Harbour de Lute forms the south-eastern side of Head Harbour Passage. The north-western side of Head Harbour Passage is formed by a series or rocks, islands, and shallow shoals.

Pervisic's (2006b) analysis uses a relatively simple synthesis of the tidal-current streams, based upon the Canadian Fisheries Department web tide model. The results suggest that the cross-sectional area of the site (Table 8.2) is 60 000 m², that the depth-averaged mean annual hydraulic power density (Chapter 3) is 1.0 kW m^{-2}, and that the resulting 'extractable' power (rated at 15 % of the total hydraulic power) is an annually averaged rate of 10 MW. This figure, however, appears to be the potential hydraulic power rather than the power output, which would amount to some 45 %, or 4.5 MW, equivalent to about 32 GW h year^{-1}.

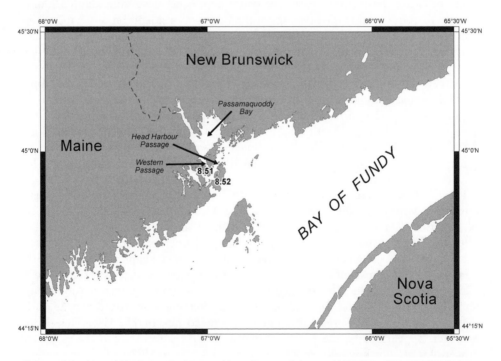

Figure 8.6   Head Harbour Passage, New Brunswick, and Western Passage, Maine.

8.5 East coast of the United States

8.5.1 Western Passage, Maine

Previsic (2006c) reports an analysis of the potential for tidal stream power in Maine, eastern USA, in which the 'Maine electricity stakeholders' selected Western Passage (Figure 8.6) because it appeared to offer the best combination of tidal resource, grid interconnection, and access to engineering support. Western Passage cuts between Moose Island on the American side and Deer Island, which is the next large island north-west of Cambabello Island, and connects Friar Roads with Passamaquoddy Bay.

Western Passage is entered between Deer Island Point, which is at the south end of Deer Island, and Dog Island, which lies off the east coast of Moose Island. It is the western conduit for waters from the St Croix River and Passamaquoddy Bay. It is said to be known for its strong currents and eddies including 'Old Sow', which is the largest whirlpool in the Western Hemisphere (Previsic 2006(c)).

There are two Admiralty (2007) diamonds in Western Passage, which here are represented by Stations 8.51 and 8.52. The first of these stations, where the Passage is wider, shows only a very low annual power output with the 'standardized' turbine. The second station, however, located at the narrowest point of the passage, shows a 'standardized' output with the STEM software of 740 MW h year^{-1} and a comparative cost of £0.203 kW h^{-1}.

Pervisic's (2006c) analysis used a simpler synthesis of the tidal-current streams, based upon the Canadian Fisheries Department web tide model. The results suggest that the cross-sectional area of the site (Table 8.2) is 17 500 m^2, that the depth-averaged mean annual hydraulic power density (Chapter 3) is 0.95 kW m^{-2}, and that the resulting 'extractable' power (rated at 15 % of the total hydraulic power) is an annually averaged rate of 2.0 MW. This figure, however, appears to be the potential hydraulic power rather than the power output, which would amount to some 45 %, or 0.9 MW, equivalent to about 7.7 GW h year^{-1}.

8.5.2 Muskeget Channel, Massachusetts

Previsic (2006d) reports an analysis of the potential for tidal stream power in Massachusetts, eastern USA. The 'Massachusetts electricity stakeholders' selected Muskeget Channel (Figure 8.7) because it appeared to offer the best combination of tidal resource, grid interconnection, and access to engineering support. Muskeget Channel is an opening 10 km wide on the south side of Nantucket Sound between Muskeget and Chappaquiddick Islands. Previsic (2006d) reports that the currents attain 3.8 knots on the flood and 3.3 knots on the ebb.

There are two Admiralty (2007) diamonds in Muskeget Channel, which here are represented by Stations 8.53 and 8.54. The first of these stations lies to the east side and exhibits only a very low annual power output with the 'standardized' turbine. The second station, however, located on the eastern side of the Channel, shows a 'standardized' output with the STEM software of 1090 MW h year^{-1} and a comparative cost of £0.138 kW h^{-1}.

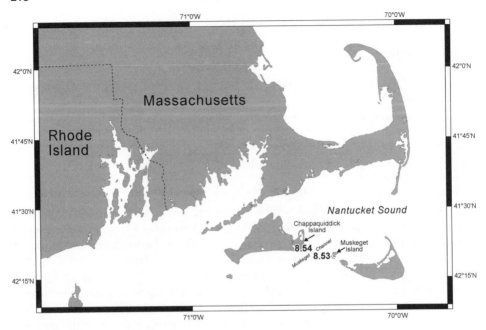

Figure 8.7 Muskeget Channel, Massachusetts.

Pervisic's (2006d) analysis used a simpler synthesis of the tidal-current streams, based upon the Canadian Fisheries Department web tide model. The results suggest that the cross-sectional area of the site (Table 8.2) is 136 000 m^2, that the depth-averaged mean annual hydraulic power density (Chapter 3) is 2.90 kW m^{-2}, and that the resulting 'extractable' power (rated at 15 % of the total hydraulic power) is an annually averaged rate of 15.6 MW. This figure, however, appears to be the potential hydraulic power rather than the power output, which would amount to some 45 %, or 7.02 MW, equivalent to about 60 GW h year^{-1}.

8.6   West coast of Canada

Station 8.55 (Figure 8.8) is located at a bay entrance near Anchorage in Alaska and shows a peak mean Spring current of less than 1 knot.

Further south, however, Station 8.56 located in Cross Sound on the north side of Chichagof Island shows a peak mean Spring current of 6.2 knots and a 'standardized' output with the STEM software of more than 2300 MW h year^{-1} at a comparative cost of £0.063 kW h^{-1}. Station 8.57 located in the narrow Peril Channel between Chicagof Island and Baranof Island shows a peak mean Spring current of 7.0 knots and a 'standardized' output from the STEM software of more than 3000 MW h year^{-1} at a comparative cost of £0.049 kW h^{-1}.

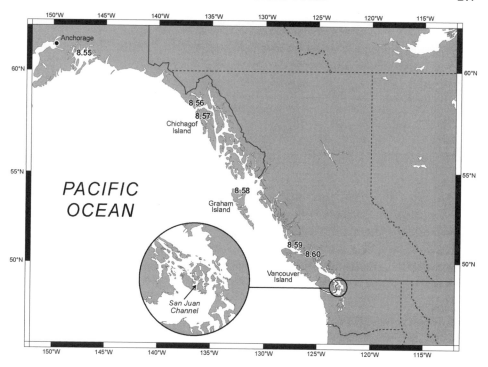

Figure 8.8 West coast of Canada.

Station 8.58 located on the north shore of Graham Island shows a peak mean Spring current of 5.5 knots and a 'standardized' output with the STEM software of 1732 MW h year^{-1} at a comparative cost of £0.087 kW h^{-1}. Station 8.59 in the channels at the northern entrance inside Vancouver Island shows a peak mean Spring current of 4.0 knots. Further south, Station 8.60 shows a peak mean Spring tide of 8.3 knots and a 'standardized' output of 4391 MW h year^{-1} at a comparative cost of £0.034 kW h^{-1}.

8.7 West coast of the United States

8.7.1 Knik Arm, Alaska

Polagye and Previsic (2006a) report an analysis of the potential for tidal stream power in Alaska, in which the 'Alaska stakeholders' selected Knik Arm (Figure 8.9) because it appeared to offer the best combination of tidal resource, grid interconnection, and access to engineering support at Anchorage. Knik Arm is located in the Upper Cook Inlet 3 km north of the city of Anchorage. Knik Arm, on either side of Cairn Point, is less than 15 m deep, but deepens to more than 50 m off Cairn Point itself. Tidal currents are strong owing to the large tidal range and the constriction of the channel of the Point.

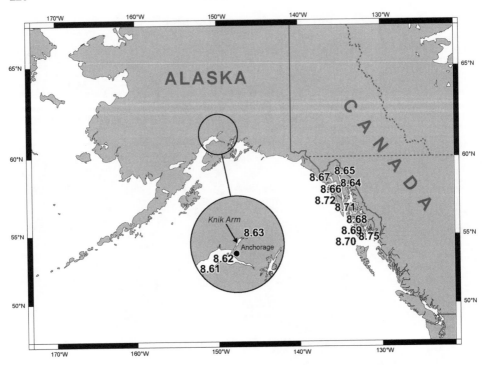

Figure 8.9 Knik Arm, Alaska.

There are three Admiralty (2007) diamonds in Cook Inlet, close to Anchorage, which here are represented by Stations 8.61 to 8.63. The first of these stations lies about 50 km to the south-west and exhibits a 'standardized' output with the STEM software of about 1100 MW h year^{-1} at a comparative cost of £0.136 kW h^{-1}. The second station, located further up the inlet, shows a similar annual power output with the 'standardized' turbine.

Polgaye and Previsics' (2006a) analysis used a simpler synthesis of the tidal-current streams, based upon the NOAA NOS data. The results suggest that the cross-sectional area of the site (Table 8.2) is 72 500 m^2, that the depth-averaged mean annual hydraulic power density (Chapter 3) is 1.60 kW m^{-2}, and that the resulting 'extractable' power (rated at 15 % of the total hydraulic power) is an annually averaged rate of 17.4 MW. This figure, however, appears to be the potential hydraulic power rather than the power output, which would amount to some 45 %, or 7.8 MW, equivalent to about 67 GW h year^{-1}.

8.7.2 South-east Alaska

Pelagye (2006d) reports the selection of the 12 sites shown in Table 8.7, based largely upon the availability of current meter data:

Table 8.7 Summary of sites in south-east Canada (data from Polagye, 2006)

Site	Station	Cross-section (m²)	Average depth (m)	Power density (kW m⁻²)	Channel power (MW)
Cross Sound and Icy Strait					
South Passage	8.64	380 000	87	1.3	480
North Passage	8.65	490 000	110	0.9	420
South Inian	8.66	34 000	46	4.3	150
North Inian	8.67	660 000	230	2.5	1600
Wrangell Narrows					
Turn Point	8.68	4700	6.8	1.8	9
South Ledge	8.69	4800	5.5	2.6	12
Spike Rock	8.70	3500	4.8	2.6	9
Chatham Strait	8.71	3100	12	7.4	23
Peril Strait	8.72	5600	11	4.5	25
Prince of Wales Island					
Tlevak Narrows	8.73	12 000	18	1.5	18
Tonowek Narrows	8.74	15 000	18	0.7	11
Felice Strait	8.75	60	1	1.6	0.3

- Cross Sound and Icy Strait: from Inian Islands east to Lemesurier;

- Wrangell Narrows: from Keene Island north to Petersburg;

- Chatham Strait: Kootznahoo Inlet;

- Peril Strait: Sergius Narrows;

- Prince of Wales Island: Tlevak and Tonowek Narrows;

- Felice Strait: Harris Island, Snipe Island, and Indian Re.

Pelagye's (2006d) analysis used a simple synthesis of the tidal-current streams, based upon the NOAA NOS tide data (Table 8.6). For comparative purposes, these have been assigned station numbers 8.64 to 8.75.

Pelagye (2006a) suggests, for example, that the cross-sectional area at the South Passage site (Table 8.2) is 380 000 m², that the depth-averaged mean annual hydraulic power density (Chapter 3) is 1.3 kW m⁻², and that the resulting 'extractable' power (rated at 15 % of the total hydraulic power) is an annually averaged rate of 480 MW. This figure, however, appears to be the potential hydraulic power rather than the power output, which would amount to some 45 %, or 7216 MW, equivalent to about 1600 GW h year⁻¹.

8.7.3 Washington and Oregon

Polagye and Previsic (2006b) also report an analysis of the potential for tidal stream power in Washington, north-western USA. The 'Washington stakeholders' selected

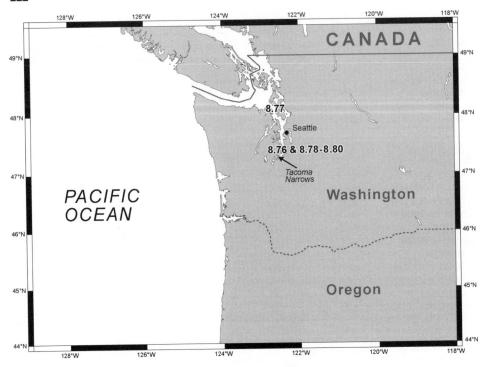

Figure 8.10 Washington and Oregon.

the Tacoma Narrows (Figure 8.10) because it appeared to offer the best combination of tidal resource, grid interconnection, and access to engineering support at Port Tacoma. The Tacoma Narrows is located in Puget Sound, 10 km west of downtown Tacoma, and separates the main basin from the south basin. While much of Puget Sound to the south and to the north of the Tacoma Narrows is deep and wide (e.g. 230 m deep and 6500 m wide between Vashon Island and the mainland), the Tacoma Narrows is relatively shallow and narrow. As a result, the semi-diurnal tides generate strong currents as the water flows through the constriction.

There is one Admiralty (2007) diamond in Takoma Narrows, which is here represented by Station 8.76, showing an annual power production of 1454 MW h^{-1} at a cost of £0.103 kW h^{-1}. Further north, Station 8.77 is in the entrance to Puget Sound and shows an annual production of 595 MW h^{-1} at a relative cost of £0.253 kW h^{-1}.

Polagye and Pervisic's (2006b) analysis used a simpler synthesis of the tidal-current streams, based upon the NOAA NOS stations, as shown in Table 8.8. The three stations are here given a STEM reference of 8.87 to 8.80.

8.7.4 California

Polagye and Bedard (2006) also report an analysis of the potential for tidal stream power in California, western USA. The 'San Francisco, California, stakeholders'

Table 8.8 Polagye and Previsic (2006b) summary of sites in the Tacoma Narrows

Station	STEM station	Depth-averaged velocity (m s^{-1}) (Spring)	P&P power (kW m^{-2})
North end mid	8.78	0.90	0.86
Port Evans	8.79	1.12	1.70
South end	8.80	1.03	1.33

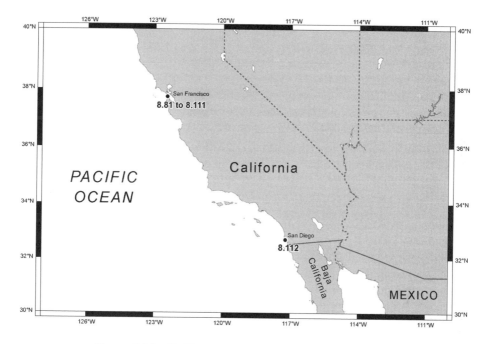

Figure 8.11 California and northern Baha California.

selected the Golden Gate (Figure 8.11) because it appeared to offer the best combination of tidal resource, grid interconnection, and access to engineering support at San Francisco. The Golden Gate Bridge spans a narrow passage that connects the San Francisco Bay to the Pacific Ocean. The tidal difference between San Francisco Bay and the open ocean forces water through this narrow channel, creating high current velocities (Table 8.9).

There is one Admiralty (2007) diamond in the Golden Gate, which here is represented by Station 8.81. The station shows a 'standardized' output with the STEM software of 1059 MW h year^{-1} at a comparative cost of £0.142 kW h^{-1}.

Polagye and Bedard's (2006) analysis used a simpler synthesis of the tidal-current streams, based upon the Canadian Fisheries Department web tide model. The results suggest that the cross-sectional area of the site (Table 8.2) is 74 700 m^2, that the depth-averaged mean annual hydraulic power density (Chapter 3) is 3.20 kW m^{-2}, and that the resulting 'extractable' power (rated at 15 % of the total hydraulic

Table 8.9 Estimates of maximum Spring and Neap tidal currents and Formzahl numbers for San Francisco Bay (from Cheng & Gardner, 1985). Also available through the book's website

STEM	C&G Ref	Spring (m s^{-1})	Neap (m s^{-1})	Spring kts	Neap kts	Formzahl number Fz	C&G M_{U2}	STEM M_{U2}	STEM S_{U2}	STEM K_{U2}
8.82	C-6	0.82	0.30	1.64	0.60	0.33	0.45	0.51	0.07	0.03
8.83	C-6	1.33	0.47	2.66	0.94	0.41	0.78	0.79	0.11	0.05
8.84	C-7	1.12	0.45	2.24	0.90	0.26	0.67	0.73	0.09	0.04
8.85	C-7	1.14	0.38	2.28	0.76	0.36	0.66	0.68	0.10	0.04
8.86	C-8	0.45	0.18	0.90	0.36	0.35	0.27	0.28	0.04	0.02
8.87	C-9	0.69	0.29	1.38	0.58	0.23	0.43	0.46	0.05	0.03
8.88	C-9	0.67	0.24	1.34	0.48	0.37	0.39	0.41	0.06	0.02
8.89	GS-9	0.96	0.37	1.92	0.74	0.26	0.59	0.62	0.08	0.03
8.90	C-10	0.50	0.23	1.00	0.46	0.23	0.32	0.34	0.04	0.02
8.91	C-10	0.53	0.25	1.06	0.50	0.31	0.34	0.36	0.04	0.02
8.92	GS-10	1.04	0.41	2.08	0.82	0.26	0.65	0.68	0.08	0.04
8.93	C11	0.47	0.22	0.94	0.44	0.24	0.31	0.32	0.03	0.02
8.94	C-12	0.83	0.30	1.66	0.60	0.26	0.50	0.53	0.07	0.03
8.95	C-12	1.17	0.50	2.34	1.00	0.31	0.74	0.76	0.09	0.04
8.96	C-13	1.02	0.42	2.04	0.84	0.24	0.64	0.68	0.08	0.04
8.97	C-13	0.87	0.33	1.74	0.66	0.31	0.54	0.55	0.07	0.03
8.98	C-14	0.65	0.28	1.30	0.56	0.31	0.40	0.43	0.05	0.02
8.99	GS-15	0.76	0.29	1.52	0.58	0.37	0.45	0.47	0.06	0.03
8.100	GS-16	0.68	0.21	1.36	0.42	0.38	0.38	0.40	0.06	0.02
8.101	GS-21	1.34	0.48	2.68	0.96	0.39	0.78	0.81	0.06	0.02
8.102	GS-22	0.97	0.39	1.94	0.78	0.33	0.59	0.62	0.06	0.02
8.103	C-225	1.43	0.57	2.86	1.14	0.31	0.85	0.91	0.06	0.02
8.104	C-226	0.57	0.23	1.14	0.46	0.44	0.33	0.35	0.06	0.02
8.105	C-302	1.05	0.43	2.10	0.86	0.31	0.66	0.68	0.06	0.02
8.106	C-304	1.29	0.53	2.58	1.06	0.40	0.78	0.80	0.06	0.02
8.107	C-306	0.95	0.39	1.90	0.78	0.38	0.59	0.60	0.06	0.02
8.108	C-307	0.46	0.19	0.92	0.38	0.35	0.27	0.29	0.06	0.02
8.109	C-310	0.57	0.27	1.14	0.54	0.36	0.38	0.38	0.06	0.02
8.110	C-312	1.05	0.50	2.10	1.00	0.40	0.40	0.69	0.06	0.02
8.111	C-313	0.39	0.16	0.78	0.32	0.30	0.25	0.25	0.06	0.02

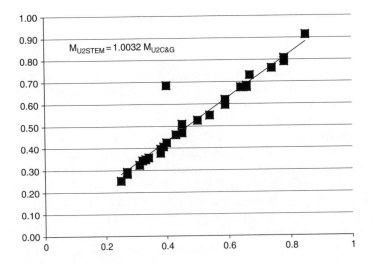

$M_{U2STEM} = 1.0032\, M_{U2C\&G}$

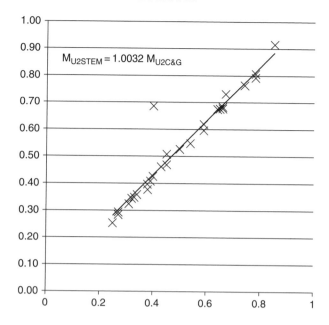

Figure 8.12 Comparison of the C&G results for M_{U2} (horizontal axis) with the STEM analysis (vertical axis) and the resulting regression equation.

power) is an annually averaged rate of 35.5 MW. This figure, however, appears to be the potential hydraulic power rather than the power output, which would amount to some 45 %, or 15.9 MW, equivalent to about 1.21 GW h year^{-1}.

There is one Admiralty (2007) diamond in the entrance to San Diego Bay, which here is represented by Station 8.112. The station shows a 'standardized' output with the STEM software of only 6 MW h year^{-1}.

Finally, the opportunity was taken to compare the magnitude of the M_{U2} harmonics computed by Cheng and Gardner (1985) with the corresponding STEM approximations based only upon the maximum mean Spring and Neap currents. The results are shown in Figure 8.12, indicating a very reasonable correspondence, even though STEM uses three input values, whereas Cheng and Gardner (1985) use 28 days of speed and direction measurements at hourly intervals.

8.8 Bibliography

Admiralty (2007) *Total Tide*. United Kingdom Hydrographic Office, Taunton, UK.

Bedard, R. (2005) *Final Survey and Characterisation: Tidal Instream Energy Conversion (TISEC) Devices*. EPRI TP-004-NA http://oceanenergy.epri.com/attachments/streamenergy/reports/ 004TISE CDeviceReportFinal111005.pdf [accessed 30 may 2008].

Bedard, R. (2006a). *North American Instream Tidal Power Study Final Briefing: East Coast.* http://www.epri.com/oceanenergy/attachments/streamenergy/briefings/05-09-06_Final_East_Coast_ Briefing_less_Env_and_Reg_Issues.pdf [accessed 30 August 2007].

Bedard, R. (2006b) *North America In-Stream Tidal Power Feasibility Study: West Coast Final Briefing.* http://oceanenergy.epri.com/attachments/streamenergy/briefings/060426_Final_West_Coast_Tidal_Briefing.pdf [accessed 30 May 2008].

Bedard, R. (2007) *EPRI Ocean Energy.* http://www.epri.com/oceanenergy/streamenergy.html [accessed 20 August 2007].

Bedard, R., Previsic, M., Polagye, B., *et al.* (2006) *North American Tidal In-Stream Energy Feasibility Study.* TP-008-NA. http://oceanenergy.epri.com/attachments/streamenergy/reports/008_Summary_Tidal_Report_06-10-06.pdf [accessed 30 May 2008].

Bedard, R. & Siddiqui, O. (2006) *Economic Assessment Methodology for Tidal In-Stream Power Plants.* EPRI-TP-02 NA Revision 2. http://archive.epri.com/oceanenergy/attachments/streamenergy/reports/002_TP_Econ_Methodology_06-10-06.pdf [accessed 2 September 2008].

CERES (2008) http://ceres.ca.gov/ceres/calweb/coastal/waters.html [accessed 2 June 2008].

Cheng, R.T., & Gartner, J.W. (1985) Harmonic analysis of tides and tidal currents in south San Francisco Bay, California. *Estuarine, Coastal and Shelf Science,* **21,** 57–74.

Cornett, A. (2007) *Inventory of Canadian Marine Renewable Energy Resources.* http://www.geomatics-atlantic.com/Files/Presentations/Andrew_Cornett.pdf [accessed 21 August 2007].

Defant, A. (1961). *Physical Oceanography.* Pergamon Press, Oxford.

DTA (2006a) *Instream Tidal Power Plant Feasibility Study: General Environmental and Federal Permitting Issues.* http://oceanenergy.epri.com/attachments/streamenergy/briefings/05-9-10-06_Final_East_Coast_Briefing_Env_and_Reg_Issues.pdf [accessed 30 May 2008].

DTA (2006b) *Instream Tidal Power in North America: Environmental and Permitting Issues.* TP-007-NA. http://oceanenergy.epri.com/attachments/streamenergy/reports/007_Env_and_Reg_Issues_Report_060906.pdf [accessed 30 May 2008].

EET (2007) *What Causes a Phytoplankton Bloom in the Gulf of Maine?* http://serc.carleton.edu/eet/phytoplankton/primer.html [accessed 20 August 2007].

EPRI (2005) *EPRI Tidal Instream Energy Conversion Feasibility Demonstration Project: Phase I Feasibility Definition Study. Kickoff Briefing.* http://oceanenergy.epri.com/attachments/streamenergy/briefings/2005TidalKickoffMeeting-TidalProject-04-20-05.pdf [accessed 30 May 2008].

Hagerman, G. (2006a) *Maine Tidal In-Stream Energy Conversion (TISEC): Survey and Characterization of Potential Project Sites.* http://oceanenergy.epri.com/attachments/streamenergy/reports/Tidal_003_ME_Site_Survey_Report_REV_1.pdf [accessed 30 May 2008].

Hagerman, G. (2006b) *Massachusetts Tidal In-Stream Energy Conversion (TISEC): Survey and Characterization of Potential Project Sites.* EPRI TP-003-MA Rev 1. http://oceanenergy.epri.com/ attachments/streamenergy/reports/Tidal_003_MA_Site_Survey_Report_REV_1.pdf [accessed 30 May 2008].

Hagerman, G. (2006c) *New Brunswick Tidal In-Stream Energy Conversion (TISEC): Survey and Characterization of Potential Project Sites.* TP-003-NB Rev 1. http://oceanenergy.epri.com/attachments/streamenergy/reports/Tidal_003_NB_Site_Survey_Report_REV_1.pdf [accessed 30 May 2008].

Hagerman, G. (2006d) *Nova Scotia Tidal In-Stream Energy Conversion (TISEC): Survey and Characterization of Potential Project Sites.* TP-003-NS Rev 1. http://oceanenergy.epri.com/attachments/streamenergy/reports/Tidal_003_NS_Site_Survey_Report_REV_2.pdf [accessed 30 May 2008].

Hagerman, G. & Polagye, B. (2006) *Methodology for Estimating Tidal Current Energy Resources and Power Production by Tidal In-Stream Energy Conversion (TISEC) Devices.* EPRI-TP-001 NA Rev 3. http://www.epri.com/oceanenergy/attachments/streamenergy/reports/TP-001_REV_3_BP_091306.pdf [accessed 30 August 2007].

International Marine (1999) *Tidal Current Tables 1999 Atlantic Coast of North America.* International Marine, ME, 209 pp.

International Marine (2006) *Tidal Current Tables 2006 Pacific Coast of North America and Asia.* International Marine, ME, 225 pp.

NCDC (2008) http://www.ncdc.noaa.gov/paleo/pubs/degaridel-thoron2005/sst-map.gif [accessed 21 August 2008].

Polagye, M. (2006) *Tidal Instream Energy Resource Assessment for South East Alaska.* EPRI-TP-003-AK. http://www.epri.com/oceanenergy/attachments/streamenergy/reports/003_TP_AK_011007.pdf [accessed 31 August 2007].

Polagye, M. & Bedard, R. (2006) *System Level Design, Performance, Cost and Economic Assessment – San Francisco Tidal Instream Power Plant.* EPRI-TP-06-SFCA. http://www.epri.com/oceanenergy/attachments/streamenergy/reports/006_CA_06-10_-06.pdf [accessed 30 August 2007].

Polagye, M. & Previsic, M. (2006a) *System Level Design, Performance, Cost and Economic Assessment – Knik Arm Alaska Tidal Instream Power Plant.* EPRI-TP-06-AK. http://www.epri.com/oceanenergy/attachments/streamenergy/reports/006-AK_06-10-06.pdf [accessed 30 August 2007].

Polagye, M. & Previsic, M. (2006b) *System Level Design, Performance, Cost and Economic Assessment – Takoma Narrows Washington Tidal Instream Power Plant.* EPRI-TP-06-MA. http://www.epri.com/oceanenergy/attachments/streamenergy/reports/TP-006-WA_Design_Feasibility_Report_010106.pdf [accessed 30 August 2007].

Previsic, M. (2006a) *System Level Design, Performance, Cost and Economic Assessment – Minas Passage Nova Scotia Tidal Instream Power Plant.* EPRI-TP-06-NS. http://www.epri.com/oceanenergy/attachments/streamenergy/reports/006_NS_RB_06-10-06.pdf [accessed 30 August 2007].

Previsic, M. (2006b) *System Level Design, Performance, Cost and Economic Assessment – New Brunswick Head Harbor Passage Tidal Instream Power Plant.* EPRI-TP-06-NS. http://oceanenergy.epri.com/attachments/streamenergy/reports/006_NB_RB_061006.pdf [accessed 30 August 2007].

Previsic, M. (2006c) *System Level Design, Performance, Cost and Economic Assessment – Tacoma Narrows Washington Tidal Instream Power Plant.* EPRI-TP-06-WA. http://www.epri.com/oceanenergy/attachments/streamenergy/reports/006_ME_RB_06-10-06.pdf [accessed 30 August 2007].

Previsic, M. (2006d) *System Level Design, Performance, Cost and Economic Assessment – Massachusetts Muskeget Channel Tidal Instream Power Plant.* EPRI-TP-06-MA. http://www.epri.com/oceanenergy/attachments/streamenergy/reports/006_ME_RB_06-10-06.pdf [accessed 30 August 2007].

Previsic, M. (2006e) *System Level Design, Performance, Cost and Economic Assessment – Maine Western Passage Tidal Instream Power Plant.* EPRI-TP-006-ME. http://oceanenergy.epri.com/attachments/streamenergy/reports/006_ME_RB_06-10-06.pdf [accessed 30 August 2007].

Previsic, M. & Bedard, R. (2005) *Methodology for Conceptual Level Design of Tidal Instream Energy Conversion (TISEC) Power Plants.* EPRI TP-005-NA. http://oceanenergy.epri.com/attachments/streamenergy/reports/005TISECSystemLevelConceptualDesignMethodologyRB08-31-05.pdf [accessed 30 May 2008].

USGS (2007) *The Chesapeake Bay: Geologic Product of Rising Sea Level* http://pubs.usgs.gov/fs/fs102-98/ [accessed 19 July 2008].

USGS (2008) *Topography, Shaded Relief, and Backscatter Intensity of the Hudson Shelf Valley, Offshore of New York.* USGS Open-File Report 03-372. http://pubs.usgs.gov/of/2003/of03-372/html/discussion.html accessed 19th November 2008.

8.9 Appendix

Table 8A.1

Ref	Adm. Ref	Lat	Long	Pk Sp kt	Pk Np kt	F_{zu}	M_2 ms⁻¹	S_2 ms⁻¹	K_2 ms⁻¹	Mean P kW	Max P kW	Output MWh/y	Cap Fact	Cost £m	£/kWh
8.1	Aransas	—	—	2.2	0.8	0.00	0.77	0.36	0.15	4	79	38	0.00	15,000	3.967
8.2	Baltimore	—	—	1.2	0.4	0.00	0.41	0.20	0.08	0	0	0	0.00	15,000	inf
8.3	Fundy	—	—	3.4	1.9	0.00	1.35	0.38	0.27	45	306	396	0.05	15,000	0.380
8.4	N York	—	—	2.3	1.6	0.00	0.99	0.18	0.20	9	98	80	0.01	15,000	1.873
8.5	Boston	—	—	1.6	1.2	0.00	0.71	0.10	0.14	0	0	0	0.00	15,000	inf
8.6	C Cod	—	—	4.8	3.7	0.00	2.17	0.28	0.43	184	908	1614	0.18	15,000	0.093
8.7	Ch'ston	—	—	3.2	1.6	0.00	1.22	0.41	0.24	33	251	293	0.03	15,000	0.514
8.8	CD Canal	—	—	2.5	1.8	0.00	1.10	0.18	0.22	16	127	141	0.02	15,000	1.067
8.9	Ch'peake	—	—	1.8	0.6	0.00	0.61	0.31	0.12	0	43	2	0.00	15,000	inf
8.10	D'ware	—	—	1.9	1.2	0.00	0.79	0.18	0.16	1	54	13	0.00	15,000	11.467
8.11	G'veston	—	—	3.8	0.8	0.00	1.17	0.77	0.23	50	389	436	0.05	15,000	0.345
8.12	E River	—	—	5.1	3.2	0.00	2.12	0.48	0.42	184	1000	1612	0.18	15,000	0.093
8.13	Key W	—	—	2.5	0.7	0.00	0.82	0.46	0.16	10	113	84	0.01	15,000	1.791
8.14	Miami	—	—	2.2	1.4	0.00	0.92	0.20	0.18	6	84	52	0.01	15,000	2.864
8.15	Mobile	—	—	2.7	0.5	0.00	0.82	0.56	0.16	13	139	118	0.01	15,000	1.278
8.16	Tampa	—	—	1.5	0.6	0.00	0.54	0.23	0.11	0	0	0	0.00	15,000	inf
8.17	Pollock	—	—	2.4	1.7	0.00	1.05	0.18	0.21	13	112	110	0.01	15,000	1.367
8.18	P'mouth	—	—	2.5	1.3	0.00	0.97	0.31	0.19	11	120	100	0.01	15,000	1.502
8.19	St Johns	—	—	2.8	1.7	0.00	1.15	0.28	0.23	23	173	198	0.02	15,000	0.760
8.20	Sa'nnah	—	—	2.7	1.8	0.00	1.15	0.23	0.23	21	157	186	0.02	15,000	0.809
8.21	Tampa	—	—	2.1	0.5	0.00	0.66	0.41	0.13	3	66	22	0.00	15,000	6.703
8.22	Tampa	—	—	2.6	0.6	0.00	0.82	0.51	0.16	11	125	100	0.01	15,000	1.506
8.23	NY Now	—	—	2.6	1.6	0.00	1.07	0.26	0.21	16	139	141	0.02	15,000	1.063
8.24	L Island	—	—	4.2	2.4	0.00	1.68	0.46	0.34	93	578	819	0.09	15,000	0.184
8.25	Throgs N	—	—	1.2	0.6	0.00	0.46	0.15	0.09	0	0	0	0.00	15,000	inf
8.26	P Rico	—	—	1.1	0.5	0.00	0.41	0.15	0.08	0	0	0	0.00	15,000	inf
8.27	Active	—	—	5.6	1.7	0.00	1.86	0.99	0.37	179	1000	1569	0.18	15,000	0.096
8.28	Adm'lty	—	—	3.7	0.7	0.00	1.12	0.77	0.22	45	357	395	0.05	15,000	0.380
8.29	Suisan	—	—	2.4	0.6	0.00	0.77	0.46	0.15	8	99	66	0.01	15,000	2.272

		Lat	Lon												
8.30	Burrard	—	—	5	1.2	0.00	1.58	0.97	0.32	122	894	1074	0.12	15,000	0.140
8.31	Carq'z	—	—	3.5	1.3	0.00	1.22	0.56	0.24	41	317	358	0.04	15,000	0.420
8.32	Decept'	—	—	7.1	4.4	0.00	2.93	0.69	0.59	404	1000	3546	0.40	15,000	0.042
8.33	G Gate	—	—	3.3	1	0.00	1.10	0.59	0.22	32	261	278	0.03	15,000	0.540
8.34	Grays	—	—	4	0.7	0.00	1.20	0.84	0.24	58	450	511	0.06	15,000	0.294
8.35	Hb Bay	—	—	1.5	0.9	0.00	0.61	0.15	0.12	0	0	0	0.00	15,000	inf
8.36	Is'ski	—	—	4.2	1.8	0.00	1.53	0.61	0.31	80	557	703	0.08	15,000	0.214
8.37	Kvichak	—	—	3.1	2.1	0.00	1.33	0.26	0.27	38	239	334	0.04	15,000	0.451
8.38	N Indian	—	—	7	1.3	0.00	2.12	1.45	0.42	285	1000	2498	0.29	15,000	0.060
8.39	Race R	—	—	3.2	1.9	0.00	1.30	0.33	0.26	38	257	333	0.04	15,000	0.451
8.40	R'mond	—	—	3.1	2.8	0.00	1.50	0.08	0.30	56	253	487	0.05	15,000	0.309
8.41	Rosario	—	—	3.1	0.7	0.00	0.97	0.61	0.19	24	212	214	0.02	15,000	0.703
8.42	S Diego	—	—	2.5	0.5	0.00	0.77	0.51	0.15	9	111	82	0.01	15,000	1.831
8.43	S Fr'isco	—	—	5.4	1.7	0.00	1.81	0.94	0.36	162	1000	1423	0.16	15,000	0.106
8.44	S'Juan	—	—	4.5	1.1	0.00	1.43	0.87	0.29	88	652	772	0.09	15,000	0.195
8.45	Sergius	—	—	7.1	2.6	0.00	2.47	1.15	0.49	321	1000	2815	0.32	15,000	0.053
8.46	Seymour	—	—	13.9	5.5	0.00	4.95	2.14	0.99	664	1000	5830	0.67	15,000	0.026
8.47	Snow P	—	—	3.7	1.3	0.00	1.28	0.61	0.26	49	373	430	0.05	15,000	0.349
8.48	Puget	—	—	5	1.4	0.00	1.63	0.92	0.33	125	903	1098	0.13	15,000	0.137
8.49	Wrangell	–	–	3.7	1	0.00	1.20	0.69	0.24	47	365	410	0.05	15,000	0.367
8.50	Minas P	45 20 N	64 24 W	4.5	2.5	0.00	1.79	0.51	0.36	114	708	1002	0.11	15,000	0.150
8.51	SN285F	44 57 N	67 02 W	2.0	1.5	0.00	0.89	0.13	0.18	3	65	30	0.00	15,000	4.947
8.52	SN285E	44 56 N	67 00 W	3.8	2.8	0.00	1.68	0.26	0.34	84	446	740	0.08	15,000	0.203
8.53	SN280Y	41 19 N	70 24 W	1.5	1.2	0.00	0.69	0.08	0.14	0	0	0	0.00	15,000	inf
8.54	SN280X	41 21 N	70 25 W	4.2	3.3	0.00	1.91	0.23	0.38	124	610	1090	0.12	15,000	0.138
8.55	SN861A	60 24 N	146 58 W	0.8	0.4	0.00	0.31	0.10	0.06	0	0	0	0.00	15,000	inf
8.56	SN864A	58 17 N	136 23 W	6.2	3.1	0.00	2.37	0.79	0.47	271	1000	2378	0.27	15,000	0.063
8.57	SN866A	57 24 N	135 37 W	7.0	3.5	0.00	2.68	0.89	0.54	349	1000	3066	0.35	15,000	0.049
8.58	SN886A	53 58 N	132 08 W	5.5	2.7	0.00	2.09	0.71	0.42	197	1000	1732	0.20	15,000	0.087
8.59	SN898B	50 34 N	126 41 W	4.0	2.0	0.00	1.53	0.51	0.31	73	490	644	0.07	15,000	0.234
8.60	SN910D	50 18 N	125 13 W	8.3	5.0	0.00	3.39	0.84	0.68	500	1000	4391	0.50	15,000	0.034
8.61	SN857C	60 43 N	151 33 W	4.8	2.2	0.00	1.79	0.66	0.36	126	838	1108	0.13	15,000	0.136

Table 8A.1 (continued)

Ref	Adm. Ref	Lat	Long	Pk Sp kt	Pk Np kt	F_{zu}	M_2 ms^{-1}	S_2 ms^{-1}	K_2 ms^{-1}	Mean P kW	Max P kW	Output MWh/y	Cap Fact	Cost £m	£/kWh
8.62	SN857B	61 00 N	151 05 W	3.9	1.9	0.00	1.48	0.51	0.30	67	453	584	0.07	15,000	0.258
8.63	SN857A	61 10 N	150 30 W	4.8	2.2	0.00	1.79	0.66	0.36	126	838	1108	0.13	15,000	0.136
8.64	S Passage	58 13 N	136 02 W			0.00									
8.65	N Passage	58 19 N	136 07 W			0.00									
8.66	S Inian	58 13 N	136 21 W			0.00									
8.67	N Inian	58 16 N	136 20 W			0.00									
8.68	Turn Pt	56 48 N	132 59 W			0.00									
8.69	S Ledge	56 37 N	132 58 W			0.00									
8.70	Spike	56 36 N	132 59 W			0.00									
8.71	Chatham	57 30 N	134 34 W			0.00									
8.72	Peril S	57 24 N	135 38 W			0.00									
8.73	Tlev Narr	55 15 N	133 07 W			0.00									
8.74	Tono Narr	55 45 N	133 20 w			0.00									
8.75	Felice S	55 00 N	131 32 W			0.00									
8.76	S918	47 18 N	122 33 W	5.2	2.5	0.00	1.96	0.69	0.39	166	1000	1454	0.17	15,000	0.103
8.77	S920	48 02 N	122 38 W	3.8	2.2	0.00	1.53	0.41	0.31	68	429	595	0.07	15,000	0.253
8.78															
8.79															
8.80															
8.81	S930	37 49 N	122 30 W	4.6	2.5	0.00	1.81	0.54	0.36	121	754	1059	0.12	15,000	0.142
8.82	C-6	–	–	1.64	0.60	0.33	0.51	0.24	0.03	0	0	0	0.00	15,000	inf
8.83	C-6	–	–	2.66	0.94	0.41	0.79	0.38	0.04	11	172	95	0.01	15,000	1.583
8.84	C-7	–	–	2.24	0.90	0.26	0.73	0.31	0.04	3	85	30	0.00	15,000	5.033
8.85	C-7	–	–	2.28	0.76	0.36	0.68	0.34	0.03	4	102	39	0.00	15,000	3.874
8.86	C-8	–	–	0.90	0.36	0.35	0.28	0.12	0.01	0	0	0	0.00	15,000	inf
8.87	C-9	–	–	1.38	0.58	0.23	0.46	0.19	0.02	0	0	0	0.00	15,000	inf
8.88	C-9	–	–	1.34	0.48	0.37	0.41	0.19	0.02	0	0	0	0.00	15,000	inf

8.89	GS-9	—	1.92	0.74	0.26	0.62	0.27	0.03	1	53	7	0.00	15,000	inf
8.90	C-10	—	1.00	0.46	0.23	0.34	0.13	0.02	0	0	0	0.00	15,000	inf
8.91	C-10	—	1.06	0.50	0.31	0.36	0.13	0.02	0	0	0	0.00	15,000	inf
8.92	GS-10	—	2.08	0.82	0.26	0.68	0.29	0.03	2	68	16	0.00	15,000	9.415
8.93	C11	—	0.94	0.44	0.24	0.32	0.12	0.02	0	0	0	0.00	15,000	inf
8.94	C-12	—	1.66	0.60	0.26	0.53	0.25	0.03	0	0	0	0.00	15,000	inf
8.95	C-12	—	2.34	1.00	0.31	0.76	0.31	0.04	5	103	44	0.01	15,000	3.397
8.96	C-13	—	2.04	0.84	0.24	0.68	0.28	0.03	1	62	13	0.00	15,000	11.928
8.97	C-13	—	1.74	0.66	0.31	0.55	0.25	0.03	0	42	1	0.00	15,000	inf
8.98	C-14	—	1.30	0.56	0.31	0.43	0.17	0.03	0	0	0	0.00	15,000	inf
8.99	GS-15	—	1.52	0.58	0.37	0.47	0.21	0.02	0	0	0	0.00	15,000	inf
8.100	GS-16	—	1.36	0.42	0.38	0.40	0.21	0.02	0	0	0	0.00	15,000	inf
8.101	GS-21	—	2.68	0.96	0.39	0.81	0.38	0.04	11	171	96	0.01	15,000	1.564
8.102	GS-22	—	1.94	0.78	0.33	0.62	0.26	0.03	1	60	10	0.00	15,000	inf
8.103	C-225	—	2.86	1.14	0.31	0.91	0.39	0.05	14	188	126	0.01	15,000	1.194
8.104	C-226	—	1.14	0.46	0.44	0.35	0.15	0.02	0	0	0	0.00	15,000	inf
8.105	C-302	—	2.10	0.86	0.31	0.68	0.28	0.03	2	75	20	0.00	15,000	7.339
8.106	C-304	—	2.58	1.06	0.40	0.80	0.34	0.04	10	155	84	0.01	15,000	1.797
8.107	C-306	—	1.90	0.78	0.38	0.60	0.25	0.03	1	60	10	0.00	15,000	inf
8.108	C-307	—	0.92	0.38	0.35	0.29	0.12	0.01	0	0	0	0.00	15,000	inf
8.109	C-310	—	1.14	0.54	0.36	0.38	0.13	0.02	0	0	0	0.00	15,000	inf
8.110	C-312	—	2.10	1.00	0.40	0.69	0.24	0.03	3	84	27	0.00	15,000	5.648
8.111	C-313	—	0.78	0.32	0.30	0.25	0.11	0.01	0	0	0	0.00	15,000	inf
8.112	S936	—	1.80	1.30	0.00	0.79	0.13	0.16	1	47	6	0.00	15,000	inf

9
Australia and New Zealand

9.1 Introduction

This chapter applies the STEM tidal stream power analysis to sites around the coastlines of Australia, which is described here and in Sections 9.2 to 9.6, and New Zealand, which is described in Sections 9.7 to 9.9

Australia is divided into the six states (taken clockwise from Cape York (Figure 9.1)) of Queensland, New South Wales, Victoria, South Australia, Western Australia, and Northern Territory. The coastline stretches for more than 25 000 km and includes the Coral Sea and Great Barrier Reef, the South Pacific Ocean, the Tasman Sea, the Bass Strait, the Great Australian Bight, the Indian Ocean, the Timor Sea, the Arafura Sea, and the Gulf of Carpentaria.

The Analysis of Tidal Stream Power   Jack Hardisty
© 2009 John Wiley & Sons, Ltd

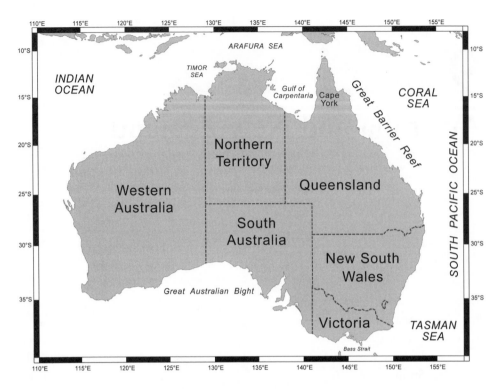

Figure 9.1 Australia and surrounding seas and oceans.

Tidal period. The frequency of the tides around Australia has been classified by the Formzahl number (Tomczak, 1998) (Chapter 4):

- $Fz < 0.25$, **semi-diurnal:** the Tasmanian coasts of the Bass Straight, the north-west coast between about Port Headland and Darwin, and a short section in the southern Coral Sea;

- $0.25 < Fz < 0.75$, **semi-diurnal with diurnal inequalities:** the east coast from Cape York to the Bass Strait, the central Great Australian Bight, and the coast between Darwin and the Gulf of Carpentaria;

- $0.75 < Fz < 2$, **mixed:** the southern coasts of Tasmania to the eastern side of the Great Australian Bight, the western side of the Great Australian Bight to the south-western corner of Australia, and the eastern and western shores of the Gulf of Carpentaria;

- $Fz_U > 2$, **diurnal:** along the remainder of the coastline.

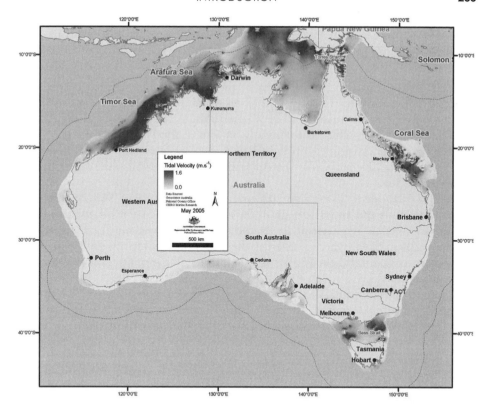

Figure 9.2 Maximum tidal currents around Australia (from CSIRO, 2008). A full colour version of this image, together with the URL of the original, is available at the book's website.

Tidal range. Spring tidal ranges are generally 1–2 m around the southern coastlines of Australia and 2–4 m around the northern coastlines. Larger ranges of 4–8 m occur in the southern Coral Sea, in the Bass Strait, to the west of Adelaide, and on the Timor coast of Western Australia. ABOM (2008) plots tidal ranges around the Australian coastlines.

Tidal currents. The map of the maximum tidal-current speeds (Figure 9.2) has been produced by the Commonwealth Scientific and Industrial Research Organization (CSIRO). The results show an overall pattern of Spring currents peaking in four areas around the Australian coastline where tidal currents in excess of 1.5 m s^{-1} occur:

(i)   the Timor Sea (Section 9.2);

(ii) the Arafura Sea (Section 9.3);

(iii) the southern region of the Coral Sea (Section 9.4);

(v)   the Bass Straits (Section 9.5).

9.2 Timor Sea

9.2.1 Geography

The Timor Sea (Figure 9.3) is an arm of the Indian Ocean situated between the island of Timor (now split between the states of Indonesia and East Timor) and the Northern Territory of Australia. The waters to the east are known as the Arafura Sea (see Section 9.3), which is technically an arm of the Pacific Ocean. The Timor Sea has two substantial inlets on the north Australian coast: the Joseph Bonaparte Gulf and the Van Diemen Gulf. The Australian city of Darwin is the only large city to adjoin the Sea.

The Sea is about 480 km wide and covers an area of about 610 000 km². The deepest section is the Timor Trough in the northern part of the sea, which reaches a depth of 3300 m. The remainder of the sea is shallower, much of it averaging less than 200 m deep, where it overlies the Sahul Shelf which is itself part of the Australian continental shelf. The sea is a major breeding ground for tropical storms and cyclones.

A number of significant islands are located in the Sea, notably Melville Island off Australia and the Australian-governed Ashmore and Cartier Islands. It is thought that early humans reached Australia by 'island-hopping' across the Timor Sea.

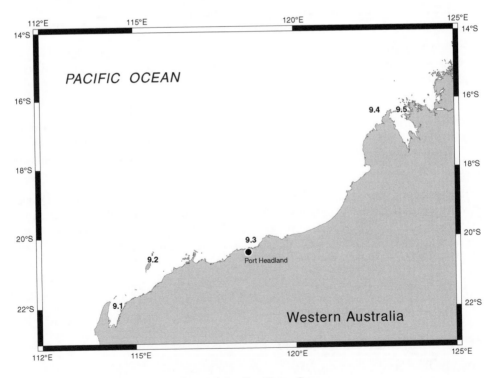

Figure 9.3 The Timor Sea.

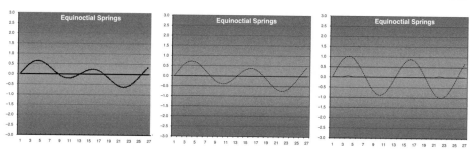

Figure 9.4 Equinoctial Spring currents for Stations 9.1 (left), 9.2 (centre), and 9.3 (right) from the STEM software and showing, in particular, the change from mixed, through semi-diurnal with inequalities, to semi-diurnal tidal regimes.

9.2.2 Oceanography

The elevation of the M_2 tide in the region is described by Robertson and Ffield (2005) and is shown in Figure 9.4. Higher amplitudes occur in the Sulawesi Sea, Indian Ocean, and shallow regions, including along the Australian coast where the magnitudes exceed 1.50 m. Lower amplitudes, <0.20 m, occur in the Java and Ceram Seas, and near an amphidromic point that forms in the Timor Sea. The M_2 tide enters this region of the Indonesian seas both from the Indian Ocean and the Pacific. In the Timor Sea, it splits, with one portion progressing into the Banda Sea and the other into the Arafura Sea. The tide progresses rapidly in the deep Banda Sea, where depths exceed 7000 m in the Weber Basin, resulting in little change in phase lag. The tide continues into the Ceram Sea, which is a confluence region, where the Indian Ocean tide meets that of the Pacific, resulting in a reduction in the tidal amplitudes.

Another confluence region occurs at the southern end of the Makassar Strait, where the tide coming through the Lombok Strait meets the Pacific tide coming through the Makassar Strait. The tide exits this region through the Java Sea. In the Java Sea, the water depth is less than 70 m, and the tide progresses slowly, resulting in closely spaced phase lines. Along the Australian coastline of the Timor Sea (Figure 9.3), the range increases from less than 2 m in the south to a maximum of more than 8 m, before decreasing to about 5 m in the north. The tidal frequencies are mainly pure semi-diurnal, but there is the presence of diurnal inequalities towards both the northern and southern coastal limits of the region (INCOIS, 2008).

9.2.3 Tidal power

There are few tidal diamonds along the Australian coast of the Timor Sea (Figures 9.3 and 9.4). In the west, Stations 9.1 to 9.3 are here allocated Fz_{UM} values of 0.6, 0.4, and 0.1 (off Port Headland) respectively in the STEM software. The stations show peak coastal Spring currents of 0.51, 0.61, and 0.97 m s^{-1} respectively, which are in agreement with the data shown in Figure 9.2. However, these flows generate less than 10 MW h year^{-1} at a rate of more than £20 kW h with the standard turbine.

Further east, the flows remain low, with peaks of 0.61 m s^{-1} at Station 9.4, which has been allocated an Fz_{UM} value of 0.15 and generates negligible tidal power with the standard turbine.

Station 9.5, however, with a similar diurnal tidal regime, evidences a peak Spring mean flow of 1.5 m s^{-1} and generates some 153 MW h year^{-1} of electricity with the standard turbine.

9.3   Arafura Sea

9.3.1   Geography

The Arafura Sea links the Timor Sea with the Pacific Ocean and overlies the continental shelf between Australia and New Guinea. It is bordered by the Torres Strait and the Coral Sea to the east (cf. Section 9.4), the Gulf of Carpentaria to the south, the Timor Sea to the west, and the Banda Sea and Ceram Sea to the north-west. It is 1290 km long and 560 km wide. The depth of the sea is primarily 50–80 m, with the depth increasing to the west. As a shallow tropical sea, its waters are a breeding ground for tropical cyclones.

The sea lies over the Arafura Shelf, which is itself part of the Sahul Shelf. When sea levels were low during the last glacial maximum, the Arafura Shelf, the Gulf of Carpentaria, and the Torres Strait formed a large flat land bridge

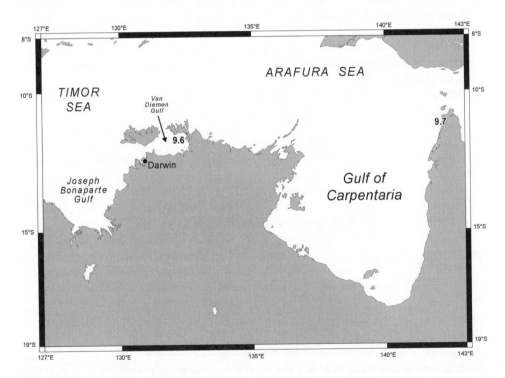

Figure 9.5 The Arafura Sea and Gulf of Carpentaria.

connecting Australia and New Guinea and easing migration of humans from Asia into Australia.

9.3.2   Oceanography

Surface salinities in the Arafura Sea range between 33.6 and 35.0 psu (Tomascik, 1997). The maximum average sea surface temperatures occur between December and February, with values of about 28.4 °C. The minimum average sea surface temperatures occur during June to August, with values of about 26.1 °C. Tidal ranges reach 6–9 m along the south coast of Java. Tidal currents are as much as 0.15 m s^{-1} off the south coast of Java and 0.25 m s^{-1} off the Australian coastline.

Webb (1981) presents a numerical model of the tides in the Gulf of Carpentaria and the Arafura Sea that is linear and time independent and uses curved boundaries. The results show that the diurnal tide has an amphidromic point near the centre of the Gulf and that the semi-diurnal tide has two amphidromic points, one in the north of the Arafura Sea and a second, virtual one, at Mornington Island. The model also shows that both frictional and resonant effects are important in determining the tides of the region.

The results show that the tidal range decreases towards the west. The Gulf of Carpentaria is essentially mixed, with diurnal characteristics on its southern shores and along a short section of its western shores. The Gulf is flanked to both the east and the west by semi-diurnal tides with diurnal inequalities. The complex pattern is also shown, for example, by the near-diurnal data for Booby Island. The results are summarized in Figure 9.6 (Admiralty, 2006).

9.3.3   Tidal power

Currents are stronger along the Australian coast of the Arafura Sea (Figure 9.2). In the region to the north-east of Darwin, for example, Station 9.6 has been allocated $Fz = 0.5$ in the present analysis and exhibits a peak coastal Spring current of around 1.94 m s^{-1}. The STEM software with the 'standard' turbine generates more than 442 MW h year^{-1} at a rate of £0.341 kW h with these hydraulic conditions.

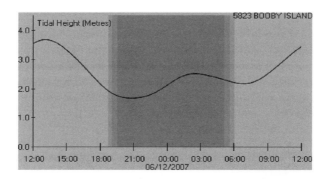

Figure 9.6 Tidal graph for Booby Island in the Torres Strait for 5 and 6 December 2007.

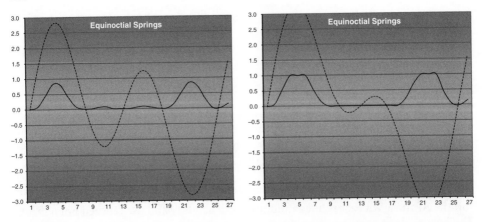

Figure 9.7 Flow and power curves from the STEM software for the Hammond Rock Lighthouse in the Torres Straits with (left) a semi-diurnal with diurnal inequalities regime and $Fz = 0.5$. Alternatively, and for a site slightly further to the west, (right) the results for a mixed regime and $Fz = 1$.

Further east, the situation becomes more interesting in the narrow Torres Straits between Australia and Indonesia. The Formzahl number shows that the tides change from semi-diurnal to mixed, diurnal dominated, as discussed above. The flows at the Hammond Rock Lighthouse (Station 9.7) (Admiralty S582) were analysed in some detail in order to explore this effect. The station was allocated $Fz = 0.5$ for a semi-diurnal with diurnal inequalities regime in the STEM software (Figure 9.7 (left)) and generated a maximum mean Spring current of $2.30\,\mathrm{m\,s^{-1}}$ and an annual energy output of 1013 MW h. Alternatively, Figure 9.7 (right) shows the effect of moving the station towards the west into a mixed tidal regime with $Fz = 1.0$. This experiment generates the same Spring flows but more than 50 % more energy, with $1613\,\mathrm{MW\,h\,year^{-1}}$.

9.4   Coral Sea

9.4.1   Geography

The Coral Sea is a marginal sea off the north-east coast of Australia (Figure 9.8) with a chain of uninhabited islands including the Willis, the Coringa, and the Tregosse Islets. It is bounded in the west by the east coast of Queensland, thereby including the Great Barrier Reef. It is bounded in the east by the New Hebrides and by New Caledonia, and in the north by the southern extremity of the Solomon Islands. The Tasman Sea lies to the south of the Coral Sea.

9.4.2   Oceanography

The surface salinity within the Coral Sea varies from about 35.0 psu in the north of the region to about 35.5 psu in the south (METOC, 2008a). The mean annual sea surface temperature decreases from about 27 °C in the north to about 24 °C in

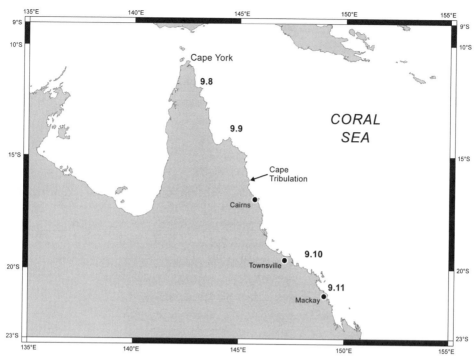

Figure 9.8 The Coral Sea.

the south (METOC, 2008b). The nearshore tides on the Australian coast of the Coral Sea are predominantly semi-diurnal with diurnal inequalities, although there is a short section around Townsville where the diurnal component becomes negligible and the regime is purely semi-diurnal. The ranges are largely of the order of 1–2 m, but become larger in the semi-diurnal section. The tidal currents follow a similar pattern, reaching maxima of less than 1 m s^{-1} for most of the coastline, but increasing to more than 1 m s^{-1} in the semi-diurnal regime area (see also Figure 9.2).

9.4.3   Tidal power

The flows north of Mackay are relatively weak. Stations 9.8 to 9.10 have here been allocated $Fz = 0.50$ and evidence peak mean Spring currents of 0.15 and 0.26 m s^{-1} respectively. This can increase, as, for example, off the headland for Station 9.9, which exhibits a peak mean Spring flow of 1.17 m s^{-1}. The STEM software generates negligible annual energy output for Stations 9.8 and 9.10 and only 53 MW h year^{-1} for Station 9.9. Offshore, in the south of the region, however, at Station 9.11 the STEM software here shows stronger flows peaking at 2.91 m s^{-1}, which generate more than 1614 MW h year^{-1} at less than £0.100 kW h^{-1}.

9.5 Bass Strait

9.5.1 Geography

The Bass Strait is a sea channel separating Tasmania from the south coast of the Australian mainland (Figure 9.9). The first European to discover it was Matthew Flinders in 1798. Flinders named it after his ship's doctor, George Bass. The Bass Strait is approximately 240 km wide at its narrowest point and generally around 50 m deep. It contains many islands, with King Island and Flinders Island home to substantial human settlements.

Like the rest of the waters surrounding Tasmania, and particularly because of its limited depth, it is notoriously rough, with many ships lost there during the nineteenth century. A lighthouse was erected on Deal Island in 1848 to assist ships in the eastern part of the Straits, but there were no guides to the western entrance until the Wilsons Promontory Lighthouse was completed in 1859, followed by another at Cape Wickham at the northern end of King Island in 1861. Strong currents are generated between the Antarctic-driven south-east portions of the Indian Ocean and the Tasman Sea's Pacific Ocean waters.

9.5.2 Oceanography

The surface salinity within the Bass Strait has average values that vary from about 35.3 psu in the south to about 35.5 psu on the Australian coast (METOC, 2008a).

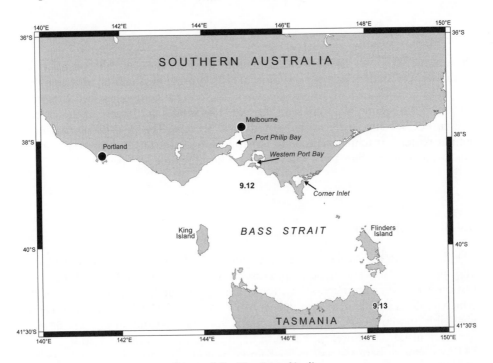

Figure 9.9 The Bass Strait.

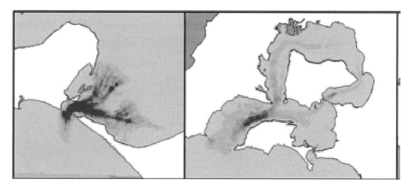

Figure 9.10 Tidal power potential in Port Phillip Bay (left) and Western Port Bay (right) on the Australian coast of the Bass Straits. Darker shading indicates increased stream power potential up to about 150 W m^{-2}.

The mean annual sea surface temperature increases from about 14 °C to about 16 °C across the same transect (METOC, 2008b). Mattom (2007) showed that the Bass Strait is mixed throughout most of the year, but becomes stratified for a few months during each summer. The shelf break front, which is established for most of the year, then makes room for shallow sea fronts on either side of the eastern and western entrance sills (Baines & Fandry, 1983).

The tidal range is 1–2 m throughout the Bass Strait, except for the central section where ranges of up to 4 m may occur. The tidal frequency is generally mixed along the south and west of the Strait and Tasmania, becoming semi-diurnal with diurnal inequalities on the north shore of the Bass Strait and towards the east. The north shore of Tasmania, however, evidences purely semi-diurnal tides.

9.5.3 Tidal power

The Victoria Government (2008) states that 'Tidal power generation opportunities along coastal Victoria are limited due to the small tidal range present in Bass Strait. However, the entrance to Port Phillip Bay and some parts of Western Port Bay present opportunities for the application of tidal energy technologies' (Figure 9.10).

Two sites were analysed with the STEM software from tidal-current diamond data in Admiralty (2006), and both were allocated $Fz = 0.5$. Station 9.12 at the entrance to Western Port bay shows a peak mean Spring current of 1.89 m s^{-1}, with which the 'standard' turbine generates 399 MW h year^{-1}. Station 9.13 in the south channel shows peaks of 1.58 m s^{-1}, with a slightly lower output.

9.6   New Zealand

9.6.1   Geography

New Zealand is an island nation in the south-western Pacific Ocean, comprising two main landmasses (the North Island and the South Island) and numerous smaller

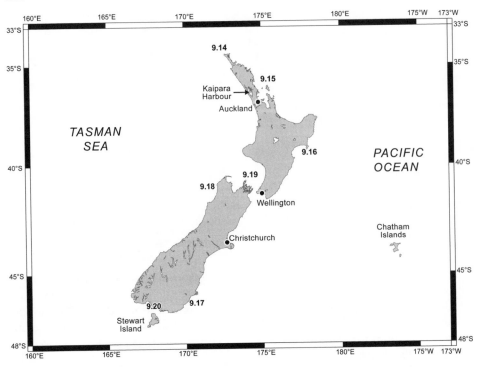

Figure 9.11   New Zealand.

islands, most notably Stewart Island and the Chatham Islands (Figure 9.11). The North and South Islands are separated by the Cook Strait, which is 20 km wide at its narrowest point. The coastline stretches for approximately 15 134 km.

The Tasman Sea is the large body of water stretching some 2000 km between Australia and New Zealand. It is a south-western segment of the South Pacific Ocean. The sea was named after the Dutch explorer Abel Janszoon Tasman, the first recorded European to encounter New Zealand and Tasmania. The British explorer Captain James Cook later extensively navigated the Tasman Sea in the 1770s as part of his first voyage of exploration. The Tasman Sea is deemed by the International Hydrographic Organization to include the waters to the east of the Australian states New South Wales, Victoria, and Tasmania. The northern state of Queensland neighbours the Coral Sea, and the boundary between New South Wales and Queensland is also used as the boundary between the two seas.

The Tasman Sea features a number of mid-sea island groups, quite apart from coastal islands located near the Australian and New Zealand coastlines. These include Lord Howe Island and its subsidiary islands, Ball's Pyramid, and Norfolk Island, in the extreme north of the Tasman Sea, on the border with the Coral Sea.

9.6.2   Oceanography

The following description of the tides around New Zealand is abstracted from Land Information New Zealand (2007). In general terms, New Zealand's tides can

Table 9.1 Tidal parameters on the west and east coasts of New Zealand

	Spring range (m)	Neap range (m)	Spring/Neap cycle
West coast	3.5–4.0	1.5–2.0	Fortnightly
East coast	1.0–2.0	0.5–1.5	Monthly

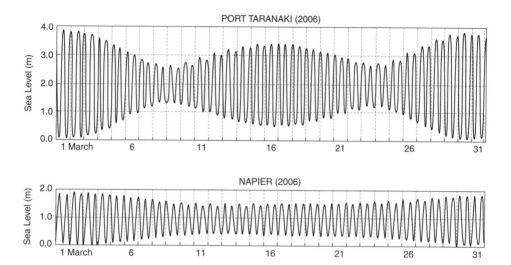

Figure 9.12 Tides at Port Taranaki (west coast, upper plot) and Napier (east coast, lower plot) for March 2008 (from Seafriends, 2008).

be grouped into two categories according to their location on either the west or east coast. The distinguishing features are the ranges and the frequencies of the Spring/Neap cycle, as shown in Table 9.1.

Figure 9.12 shows the predicted tidal curves at Port Taranaki (west coast) and Napier (east coast). Port Taranaki exhibits the classic fortnightly Spring/Neap cycle caused by the interaction of the M_2 and S_2 components.

The diurnal inequality, where successive high or low water heights are not the same, is also apparent in the diagrams. This phenomenon is particularly evident in the Port Taranaki example for high waters during the periods 8–16 and 22–30 March.

The tidal wave travels around New Zealand's coastline in an anticlockwise direction in about 13 h, corresponding to a semi-diurnal cycle (Seafriends, 2008). Between the North and South Islands, the tide travelling northwards along the east coast passes the southbound tide to the west. The strong difference between the states of these tides, combined with the east/west tide height difference, produces the complex pattern of strong currents that flow through the Cook Strait.

9.6.3   Tidal power

All of the New Zealand stations have been allocated $Fz = 0$ corresponding to a semi-diurnal tidal regime. The coastal currents down the east coast of both North Island and South Island are relatively weak, Stations 9.15, 9.16, and 9.17 showing peaks of 0.31, 0.15, and 0.26 m s^{-1} respectively and low potential for power outputs. Currents are a little stronger off the east coast, with, for example, Station 9.18 peaking at only 0.56 m s^{-1}.

Currents are stronger off the northern cape of North Island, in the Cook Strait, and off the southern cape of South Island. Thus, Stations 9.14, 9.19, and 9.20 show peaks of 1.48, 1.22, and 2.19 m s^{-1} respectively. The STEM software calculates that the corresponding energy outputs with the 'standard' turbine are 195, 69, and 957 MW h year^{-1} respectively.

It is in some of New Zealand's enclosed bays, however, that the greatest potential for tidal power exists. For example, Crest Energy Limited (2007) has applied for consent to construct a marine turbine power generation project in the Kaipara Harbour in Northland, northern New Zealand. The project comprises up to 200 completely submerged marine tidal turbines with a maximum generating capacity of around 200 MW, located near the entrance of the Harbour. Crest Energy estimate its plans will generate energy for up to 250 000 NZ homes by harnessing the power of the tidal flows into and out of the Kaipara Harbour. The project may contribute 3 % of New Zealand's supply. The harbour is one of the largest harbours in the world, covering 900 km^2, with 3000 km of shoreline. The Kaipara extends for 60 km north to south.

In a separate development, Renewable Energy Development (2008) reported that a prototype, 14 m diameter, carbon-fibre, tidal turbine is being prepared for a trial in Cook Strait. The device is reported to be for installation in the 'Karori Rip', 4.5 km off Wellington's Island.

9.7   Bibliography

ABOM (2008) *Tide Predictions for Australia, South Pacific and Antarctica*. Australian Bureau of Meteorology. http://www.bom.gov.au/oceanography/tides/index_range.shtml [accessed 19 June 2008].

Admiralty (2006) *Total Tide*. United Kingdom Hydrographic Office, Taunton, UK.

Baines, P.G. & Fandry, C.B. (1983). Annual cycle of the density field in Bass Strait. *Aust. J. Mar. Freshw. Res.* 34: 143–53.

Crest Energy Ltd. (2007) http://www.crest-energy.com/company.htm# [accessed 27 November 2007].

CSIRO (2008). http://www.marine.csiro.au/datacentre/noo_public/ocean_2004/maps/tides/max_tidal_vel.jpg [accessed 21 August 2008].

INCOIS (2008) http://www.incois.gov.in/Tutor/ShelfCoast/chapter18.html [accessed 18 June 2008].

Land Information New Zealand (2007) *Tides Around New Zealand*. http://www.linz.govt.nz/home/index.html [accessed 27 November 2007].

Mattom (2007) *Lecture Notes on Coastal Fronts*. http://www.es.flinders.edu.au/~mattom/ShelfCoast/chapter09.html [accessed 24 November 2007].

METOC (2008a) www.metoc.gov.au/products/data/aussss.php [accessed 18 June 2008].

METOC (2008b) http://www.metoc.gov.au/products/data/aussst.php [accessed 18 June 2008].

Renewable Energy Development (2008) *Tidal Energy from Cook Strait*. http://renewableenergydev.com/red/tidal-energy-cook-strait/ [accessed 19 June 2008].

Robertson, R. & Ffield, A. (2005) M_2 baroclinic tides in the Indonesian Seas. *Oceanography* **18**, 4. http://www.ldeo.columbia.edu/~rroberts/18.4_robertson_ffield_hi.pdf [accessed 25 November 2007].

Seafriends (2008) http://www.seafriends.org.nz/oceano/tides2.gif [accessed 19 June 2008].

Tomascik, T. (1997) *The Ecology of the Indonesian Seas.* Periplus, Hong Kong.

Tomczak (1998) http://www.incois.gov.in/Tutor/ShelfCoast/chapter18.html accessed 19th November 2008.

Victoria Government (2008) http://www.sustainability.vic.gov.au/www/html/2118-tidal.asp [accessed 18 June 2008].

Webb, D.J. (1981) Numerical modal of the tides in the Gulf of Carpentaria and the Arafura Sea. *Australian Journal of Marine and Freshwater Research* **32**(1), 31–44.

9.8   Appendix: STEM outputs for Australia and New Zealand

Table 9A.1   STEM outputs for Australia and New Zealand

Ref	Adm. Ref	Lat	Long	Peak Spring (knots)	Peak Neap (knots)	FZ_u	M_2 (m s^{-1})	S_2 (m s^{-1})	K_2 (m s^{-1})	Mean P (kW)	Max P (kW)	Output MWh year^{-1}	Cap Fact	Cost £ million	£ kW h^{-1}
9.1	SN6225L	22 21 S	114 14 E	1.0	0.3	0.60	0.265	0.143	0.013	0	0	0	0	15000	inf
9.2	SN625Q	20 18 S	115 31 E	1.2	0.4	0.40	0.354	0.177	0.018	0	0	0	0	15000	inf
9.3	SN626K	20 13 S	118 32 E	1.9	0.4	0.10	0.567	0.37	0.028	0.128	41.69	1.124	1E-04	15000	inf
9.4	SN627F	16 46 S	122 09 E	1.2	0.3	0.15	0.363	0.218	0.018	0	0	0	0	15000	inf
9.5	SN628A	16 11 S	123 06 E	3.0	1.2	0.15	1.017	0.436	0.051	18.57	176.1	162.9	0.019	15000	0.9229
9.6	SN633A	12 05 S	131 05 E	3.8	1.6	0.50	1.148	0.468	0.057	50.33	558	441.6	0.05	15000	0.3405
9.7	S582	10 30 S	142 13 E	4.5	3.3	0.50	1.658	0.255	0.083	115.5	932.7	1013	0.116	15000	0.1484
9.8	SN587B	12 31 S	143 38 E	0.3	0.1	0.50	0.085	0.043	0.004	0	0	0	0	15000	inf
9.9	SN588A	13 58 S	144 25 E	2.3	0.7	0.50	0.638	0.34	0.032	5.984	123.4	52.51	0.006	15000	2.8637
9.10	SN953A	19 16 S	147 30 E	0.5	0.2	0.50	0.149	0.064	0.007	0	0	0	0	15000	inf
9.11	SN593F	19 55 S	150 17 E	5.7	2.6	0.10	2.046	0.764	0.102	183.9	1000	1614	0.184	15000	0.0932
9.12	SN607C	38 27 S	145 10 E	3.7	1.5	0.50	1.105	0.468	0.055	45.47	514.9	399	0.046	15000	0.3769
9.13	SN611B	40 39 S	148 05 E	3.1	2.5	0.50	1.19	0.128	0.06	35.76	305.4	313.8	0.036	15000	0.4792
9.14	SN638H	34 23 S	172 40 E	2.9	1.7	0.00	1.173	0.306	0.059	22.24	138.1	195.1	0.022	15000	0.7706
9.15	SN639P	35 10 S	174 22 E	0.6	0.4	0.00	0.255	0.051	0.013	0	0	0	0	15000	inf
9.16	SN643B	39 31 S	178 00 E	0.3	0.2	0.00	0.128	0.026	0.006	0	0	0	0	15000	inf
9.17	SN650A	45 47 S	170 50 E	0.5	0.4	0.00	0.23	0.026	0.011	0	0	0	0	15000	inf
9.18	SN652A	40 44 S	172 13 E	1.1	0.6	0.00	0.434	0.128	0.022	0	0	0	0	15000	inf
9.19	SN644G	41 13 S	174 29 E	2.4	1.2	0.00	0.918	0.306	0.046	7.816	77.79	68.58	0.008	15000	2.1927
9.20	SN650AS	46 36 S	168 21 E	4.3	3.0	0.00	1.862	0.332	0.093	109.1	453.7	957.1	0.109	15000	0.1571

10
Rest of the world

10.1 Introduction

This chapter considers the potential for tidal stream power at sites throughout the rest of the world. The global distributions of tidal parameters that apply were detailed in Chapter 5 and are summarized below.

Global tidal ranges are essentially controlled by distance from the main amphidromic points which form as the global ocean oscillates in response to the tide-generating forces (Chapters 1 and 2). There are twelve major oceanic amphidromic points:

1. North Atlantic.

2. Bahamian.

3. South Atlantic.

4. South Atlantic.

5. Madagascan.

6. South-west Australian.

7. North-west Pacific.

8. North-east Pacific.

9. Arctic.

10. Central Pacific.

11. East-central Pacific.

12. South Pacific.

Global tidal currents. These tidal systems lead to tidal ranges that, in general, increase with distance from the amphidromic points, and then to a tidal-current system that reflects the tidal ranges but depends upon local resonances and coastal and sea floor topographies. There are six regions that exhibit strong tidal currents but are not covered in previous chapters. These regions are dealt with in the present chapter:

1. The Barents Sea (Section 10.2).

2. The east coast of South America (Section 10.3).

3. The western Indian Ocean (Section 10.4).

4. The East Arabian Sea (Section 10.5).

5. Indonesia and the Java Sea (Section 10.6).

6. The Sea of Japan (Section 10.7).

The geography and the oceanography of these regions is described below.

Tidal power. The potential tidal power that is available in each of these regions depends, as discussed in previous chapters, on a combination of the tidal-current harmonics and the tidal regime, as specified by the Formzahl number. The *simplified tidal economic model* (STEM) developed in Chapter 5 was set up with 'standardized' inputs for the analyses in this chapter. The site-specific, locational, and flow parameters were generally taken from Admiralty (2005) and from the Formzahl chart. The software was then utilized, alongside some recent research reported in the literature, to investigate the potential for, and relative cost of, tidal stream power developments in these regions.

The conclusion has to be that there are areas in the rest of the world with vast potential for the exploitation of tidal stream power. These include the Gulfs of

Kutch and Cambay on the Indian west coast, certain channels around New Zealand and Australia, in Indonesia, and along the coastlines around China and Korea. The developments may, however, be hindered by the costs of the early-stage technology and the installation and grid connection problems. This global analysis has also highlighted the requirement for the type of full-year hydrodynamic syntheses that are modelled by STEM. In particular, it is frequently observed that two sites with similar peak Neap and Spring currents can generate quite different electrical energy outputs as a function of the transition from semi-diurnal to diurnal tidal regimes.

10.2 Barents Sea

10.2.1 Geography

The Barents Sea (Figure 10.1) is a part of the Arctic Ocean located north of Norway and of Russia. It is a shelf sea with an average depth of 230 m, bordered by the shelf edge towards the Norwegian Sea in the west, the island of Svalbard (Norway) in the north-west, and the islands of Franz Josef Land and Novaya Zemlya in the north-east and east. Novaya Zemlya separates the Kara Sea from the Barents Sea.

Known in the Middle Ages as the *Murman Sea*, it takes its modern name from the Dutch navigator Willem Barents. The southern half of the Barents Sea, including the ports of Murmansk (Russia) and Vardø (Norway), remain ice-free year round owing to the warm North Atlantic drift. In September, the entire Barents Sea is more or less completely ice-free.

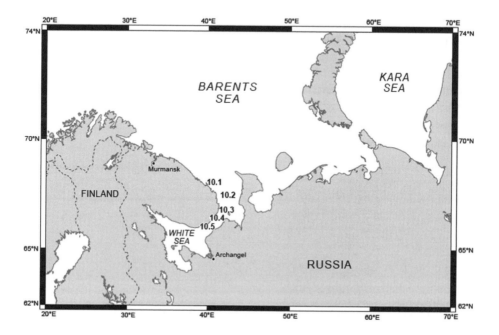

Figure 10.1 The Barents Sea.

10.2.2 Oceanography

There are three main types of water masses in the Barents Sea:

(a) warm, saline Atlantic water (temperature >3 °C, salinity >35 psu) from the North Atlantic drift;

(b) cold Arctic water (temperature <0 °C, salinity <35 psu) from the north;

(c) warm, low-salinity coastal water (temperature >3 °C, salinity <34.7 psu).

Between the Atlantic and the Polar waters, a convergence known as the Polar Front is formed. In the western parts of the Barents Sea, this front is determined by the bottom topography and is therefore relatively sharp and stable from year to year. In the eastern parts of the Sea it can be quite diffuse, and its position can vary from year to year.

There are regions exhibiting strong tidal flows, in excess of $3 \, \mathrm{m \, s^{-1}}$, in the Arctic Ocean. In certain areas, such as the Barents Sea, vertical mixing is enhanced by the dissipation of the tidal energy through bottom friction. The consequent upward mixing contributes to the melting of the sea ice melt. Tides also create cracking in the sea ice, which allows ocean heat to escape and also mobilizes the pack.

Tidal ranges are large in the Barents Sea itself (Hollowplanets, 2008), increasing to amplitudes of more than 2.0 m in the southern coastal regions. The tides are driven by a semi-diurnal amphidrome located in the north-western sector of the Sea and are probably a resonant response to the ocean forcing (Gjevik, Nøst, & Straume, 1994). The maximum tidal currents in the Arctic Ocean are due to M_2, S_2, O_1, and K_1 tidal constituents and are described by Kowalik, Proshutinsky, and Thomas (2007). The largest currents in the Arctic Ocean occur at the entrance to the White Sea and at the Spitsbergenbanken, which lies south of Spitsbergen (Kowalik & Proshutinsky, 1993). Topographically amplified semi-diurnal tidal currents are found in the vicinity of Bear Island in the Barents Sea (Kowalik, 1994; Kowalik & Proshutinsky, 1995).

10.2.3 Tidal power

The strongest tidal currents occur in the Barents Sea at the narrowing entrance to the White Sea. The area was analysed with STEM software, and all stations were allocated $Fz = 0$, as detailed in earlier chapters.

The analysis utilized the Admiralty (2005) data along a north – south line at the entrance to the White Sea with Stations 10.1 to 10.5. The results are shown in Figure 10.2.

The mean peak Spring currents are 1.5, 2.8, 2.4, 3.3, and $1.3 \, \mathrm{m \, s^{-1}}$ respectively. Thus, the currents rise to a peak around the main change in direction of the Channel at Station 10.4.

The STEM software was utilized with the 'standard' turbine (see Chapter 5) at Station 10.4. The site generated more than $360 \, \mathrm{MW \, h \, year^{-1}}$ at a rate of £0.414 kW h.

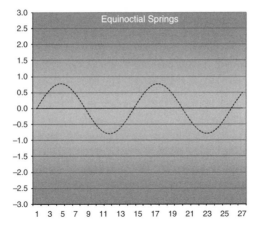

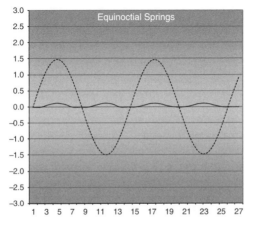

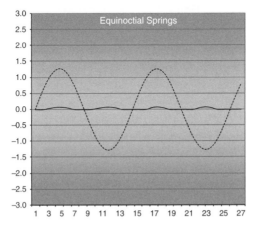

Figure 10.2 Equinoctial tidal-current speed and electrical output for Stations 10.1 to 10.5 (upper to lower) in a north to south transect from the Barents Sea to the White Sea.

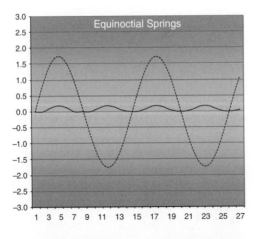

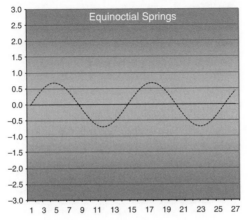

Figure 10.2 (Continued)

Thus, the prospects for tidal stream energy generation are reasonable in the channels between the Barents Sea and the White Sea, but, at present, there are many, significantly less expensive, locations in the rest of the world.

10.3 Western South Atlantic

10.3.1 Geography

There are three regions in the western South Atlantic within which there appear to be strong tidal currents: the Amazon Shelf, the Rio de Janeiro Shelf, and the Falkland Islands.

10.3.1.1 Amazon Shelf

The Amazon Shelf is, of course, dominated by the Amazon River, which is the largest river in the world by volume, with a total river flow greater than the total ocean

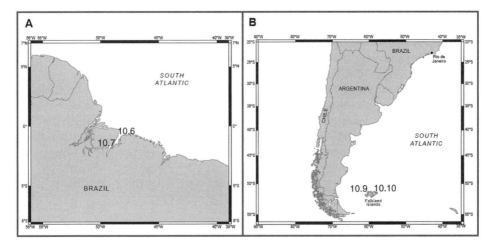

Figure 10.3 (Left) General features of the Amazon Shelf and (right) the Falkland Islands.

input of all of the next top 10 largest rivers, and accounts for approximately one-fifth of the world's total river flow. The mouth of the Amazon estuary is more than 330 km wide, which is more than the entire length of the River Thames in England. The Amazon Shelf is dominated by the river inputs, and is bisected by the submerged drainage channels minating from the river's mouth (Figure 10.3 (left)).

10.3.1.2 Rio de Janeiro Shelf

The Rio de Janeiro Shelf runs from north of the Brazilian city in the north to beyond Monte Video on the Rio de la Plata estuary in Uraguay to the south (Figure 10.3 (right)). It is a relatively narrow continental-shelf region, but widens towards the south.

10.3.1.3 Falkland Islands

The Falkland Islands (Spanish: *Islas Malvinas*) are an archipelago in the South Atlantic Ocean, located 483 km east from the coast of Argentina, 1080 km west of South Georgia, and 940 km north of Elephant Island in Antarctica (Figure 10.4 (right)). The Falkland Islands consist of two main islands, East Falkland and West Falkland, together with 776 smaller islands. Stanley, on East Falkland, is the capital city.

10.3.2 Oceanography

The mean August sea surface temperatures in the South Atlantic are described by Schweitzer (1993). The sea surface temperature decreases from about 30 °C on the north coast of South America to 25 °C at the southern extremity of the Amazon Shelf, 20 °C off Rio de Janeiro, and about 5 °C at the latitude of the Falkland Islands.

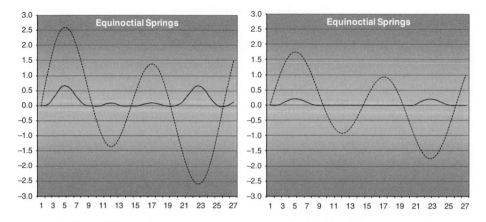

Figure 10.4 Equinoctial tide STEM outputs for (left) Station 10.6 and (right) Station 10.7.

The corresponding sea surface salinities have a value of about 34.5 psu at the mouth of the Amazon River and increase to more than 35 psu off Rio de Janeiro (University of Minnesota, 2005). The values then decrease in a southerly direction to 34.5 psu off the coast of Argentina and around the Falkland Islands. The tidal regimes are detailed in the following sections.

10.3.2.1 Amazon Shelf

The Amazon shelf is a dynamic region, dominated by the outflow from the Amazon River. Mean annual water discharge from the Amazon into the Atlantic Ocean is approximately $1.8 \times 10^5 \, m^3 \, s^{-1}$. Amazon River water can migrate as far north as Barbados and out to a distance of more than 300 km from the coastline. The plume can evidence changes on a scale of days, which can be larger than the seasonal variations. Tidally averaged cross-shelf flow has a landward component of $0.05–0.10 \, m \, s^{-1}$ within the plume's salt wedge, and a seaward flow of $0.4–1.0 \, m \, s^{-1}$ in the overlying brackish plume. However, cross-shelf tidal velocities can reach $1–2 \, m \, s^{-1}$. The North Brazil current also reaches speeds of $1–2 \, m \, s^{-1}$ and forces Amazon River water and sediments to the north-west.

10.3.2.2 Rio de Janeiro Shelf

de Mesquita and Harari (2007) report that the semi-diurnal tidal constituents (M_2, N_2, K_2, and S_2) propagate in an anticlockwise direction north of Rio but in a clockwise direction between Rio and Paranaguá. Thereafter, the amplitudes of these semi-diurnal components show a general increase from north to south. However, the diurnal components Q_1 and O_1 propagate in a clockwise direction, while K_1 and P_1 follow the semi-diurnal components in an anticlockwise sense. Such opposite senses of propagation suggest that the K_1 has two amphidromes in the South Atlantic, one propagating clockwise and the other in an anticlockwise direction.

10.3.2.3 Falkland Islands

Offshore, sea surface temperatures range from around 6 °C in July/August to 13 °C in February. Inshore sea temperatures range from 2 to 14 °C in the same period (Falkland Islands Government, 2007). Sea bottom temperatures at 250 m depth range from 5 °C in winter to 6 or 7 °C in summer. To the north of the Islands, the Falklands current has a mean velocity of about 0.25 m s^{-1} towards the north-east, but strong currents can be expected at times owing to the unsettled and stormy characteristics of the region. To the south-west and east of the Islands, the Southern Ocean current has a mean velocity of about 0.25–0.50 m s^{-1} towards the north and north-east. The tides around the Islands are semi-diurnal, with high waters from 0.3 to 3.5 m above local datum.

10.3.3 Tidal power

10.3.3.1 Amazon Shelf

As suggested above, there are strong tidal currents associated with the mouth of the Amazon River, which are due, in part, to the vast volume of water that enters and exits the river in each tide. For example, Station 10.6 represents currents at the mouth of the Amazon and has been allocated $Fz = 0.4$ for the present analysis (Figure 10.4). Application of the data to the STEM software shows peak mean Spring currents of 2.19 m s^{-1}, which with the 'standard' turbine generates 646 MW h year^{-1} at a rate of £0.233 kW h^{-1}. Currents drop off upstream, and Station 10.7 shows peaks of 1.48 m s^{-1} and only 186 MW h year^{-1} at a cost of £0.809 kW h^{-1}.

Goreau *et al.* (2005) state that a project in a tidal stream near the mouth of the Amazon River in Brazil generates power for a small isolated community. The turbine and power system were built by villagers in local workshops, with the only imported item being the turbine blades. The cost of the materials and equipment is considerably less than equivalent photovoltaic panels, and maintenance and operational costs are reported to be less than those of diesel generators.

10.3.3.2 Rio de Janeiro Shelf

The only Admiralty tidal-current data in this region is to the south of Santos in the entrance to Paranagua. Station 10.8 was allocated $Fz = 0.0$, which signifies pure semi-diurnal conditions in these analyses. The STEM software indicated a peak mean Spring current of 2.24 m s^{-1}, which is capable of generating a similar output to that given for the mouth of the Amazon above.

10.3.3.3 Falkland Islands

The strongest tidal currents listed by the Admiralty occur at the western cape of West Falklands. Station 10.9 was allocated $Fz = 0.5$, on the margins between pure semi-diurnal and mixed with semi-diurnal dominance. The STEM software with these data produced a peak mean Spring current of 2.09 m s^{-1}, which was capable of generating 622 MW h year^{-1} at a rate of about £0.242 kW h^{-1}. Although this is a not insignificant amount of electrical energy for a local community, the cabling costs from western West Falklands to the only major settlement at Stanley on East Falkland would be high, and localized wind turbines may therefore be more effective. Station 10.10 yielded low flow speeds and energy generation potential.

10.4   Western Indian Ocean

10.4.1   Geography

The western Indian Ocean (WIO) encompasses a wide range of environmental settings and extends from Somalia in the north to the subtropical waters off the coast of northern South Africa (Figure 10.5). The region includes the waters and coastal zones of Somalia, Kenya, Tanzania, Mozambique, and South Africa, as well as the island states of Mauritius, Comoros, Seychelles, Reunion and Madagascar. The region extends from latitude 0° to 30° South and longitude 30° to 65° East.

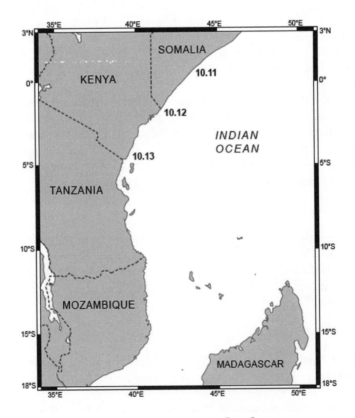

Figure 10.5 The western Indian Ocean.

10.4.2   Oceanography

The sea surface temperatures off the eastern coastline of Africa are described by Schweitzer (1993). The sea surface temperatures are about 25 °C and decrease by a few degrees towards the south of the region. The south equatorial current and the east African coastal current run offshore and are strongest during the south-west monsoon. The east Madagascar currents and the Mozambique current systems, however, are strongest during the north-east monsoon. The tidal ranges in the region show considerable variation. The microtidal areas include Mauritius and the Reunion Islands, the mesotidal areas include the Seychelles and the east coast of Madagascar, while the macrotidal areas include the mainland coasts, the Comoros, and the west coast of Madagascar.

10.4.3   Tidal power

Although the data shown earlier suggest a good potential in the western Indian Ocean, this potential does not appear to be supported by the data available. Reasonably strong semi-diurnal tidal currents are listed by the Admiralty at the entrance to bays on the Somalia/Kenya border. For example, Station 10.11 was allocated $Fz = 1.5$ and showed a peak mean Spring current of $0.71\,\mathrm{m\,s^{-1}}$ with little tidal power generation potential. Coastal currents are also weak: Stations 10.12 and 10.13 show mean peak Spring currents of little more than 0.50–$0.60\,\mathrm{m\,s^{-1}}$.

10.5   East Arabian Sea

10.5.1   Geography

The Arabian Sea (Figure 10.6) is a region of the Indian Ocean bounded on the east by India, on the north by Pakistan and Iran, on the west by the Arabian Peninsula, and on the south by a line between Cape Guardafui at the north-east point of Somalia and Cape Comorin in India. The maximum width of the Arabian Sea is approximately 2400 km, and its maximum depth is 4652 m in the Arabian Basin.

The Arabian Sea has two important branches: the Gulf of Aden in the south-west, connecting with the Red Sea through the strait of Bab-el-Mandeb, and the Gulf of Oman to the north-west, connecting with the Persian Gulf. There are also two embayments on the Indian coast, known as the Gulf of Cambay and the Gulf of Kutch, which are of great importance here and are detailed below.

The Gulf of Cambay (also known as the Gulf of Khambhat) is an inlet of the Arabian Sea in the state of Gujarat on the west coast of India. The Gulf is about 150 km in length and divides the Kathiawar peninsula to the west from the eastern part of Gujarat state to the east. The Narmada River and Tapti River empty into the Gulf. The Gulf is shallow and exhibits many shoals and sandbanks. The Gulf is known for the amplitude and phase speed of its extreme tides (Nayak & Shetye, 2003) (Figure 10.7). The Alang Ship Recycling Yard takes advantage of the macrotidal range in the Gulf. Large ships are beached during the twice-monthly highest tides, and are dismantled when the tide recedes. The Gulf of Kutch is an inlet of the

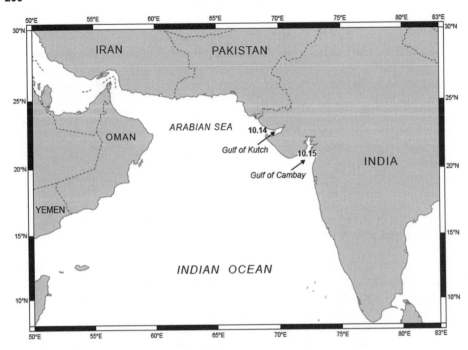

Figure 10.6 The Arabian Sea.

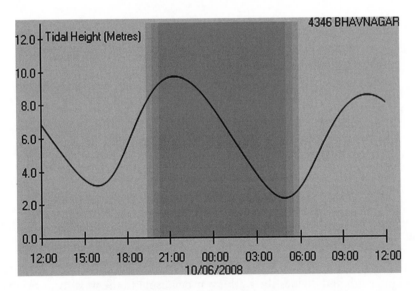

Figure 10.7 Tidal curve for Admiralty reference 4346 Bavinhagar in the Gulf of Cambay for 080609 and 080610 (from Admiralty, 2005).

Arabian Sea, also in the state of Gujarat on the west coast of India. It is about 150 km in length and divides Kutch and the Kathiawar peninsula regions of Gujarat. The Rukmavati River empties into the Arabian Sea nearby.

10.5.2 Oceanography

The sea surface temperatures off the western coastline of India are described by Schweitzer (1993). The sea surface temperatures are around 28 °C in the north and increase by a few degrees towards the south of the region. The corresponding sea surface salinities decrease from about 35 psu in the north of the region to about 33 psu in the south (University of Minnesota, 2005).

The M_2 tides in the East Arabian Sea are controlled by the Madagascan amphidromic point (MMSTATE, 2007a). In general, the M_2 amplitude increases in a northerly direction along the west coast of India between 0.05 to 0.10 m in the south and 0.60 to 0.70 m in the north. In the Gulf of Cambay (Figure 10.8),

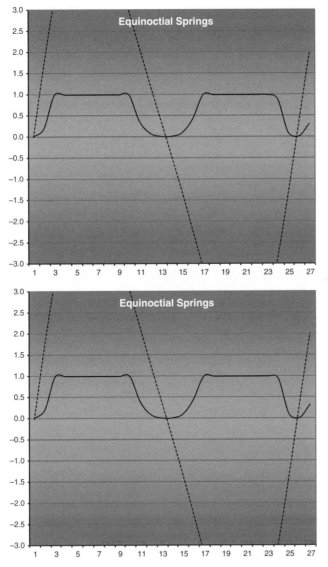

Figure 10.8 Equinoctial tide STEM outputs for (left) Station 10.14 and (right) Station 10.15.

however, local resonances are responsible for increasing the M_2 amplitude to more than 3.00 m (Nayak & Shetye, 2003).

The tides in the Arabian Sea exhibit a mixed nature and change from predominantly diurnal in the south-west quadrant to predominantly semi-diurnal in the north-east. The tides tend to become more semi-diurnal in nature as shallower depths are approached, such as the continental shelf. It is also in these shelf areas that the compound tides become more pronounced and thus may play an increasingly important role The Formzahl number exhibits rapid coastwise change along the Arabian Sea margins of eastern India. Figure 10.8 shows that Fz_{MU} varies from about 1.00 (mixed) to in excess of 2.00 (diurnal) and is also particularly high in the Gulf of Cambay.

10.5.3   Tidal Power

The strongest tidal currents listed by the Admiralty occur in the two Indian Gulfs and are illustrated by Stations 10.14 and 10.15. $Fz = 2.0$ is appropriate for the analyses of these stations with the STEM software.

Station 10.14 (Figure 10.8) shows the results in the entrance to the Gulf of Kutch, with a peak mean Spring current of 2.55 m s^{-1} capable of generating more than 3792 MW h year^{-1} at a rate of about £0.040 kW h^{-1}.

Station 10.15 shows the results in the entrance to the Gulf of Cambay, with a peak mean Spring current of 2.60 m s^{-1} capable of generating a similar amount of electricity and at a similar cost.

There is clearly a very significant potential for tidal stream power generation in these two Gulfs on the west coast of India, local construction costs may well be reasonable, and the country as a whole has a great demand for low-cost power.

10.6   Indonesia and the Java Sea

10.6.1   Geography

The Republic of Indonesia is a nation in south-east Asia comprising 17 508 islands; it is the world's largest archipelagic state. The nation's capital city is Jakarta. The country shares land borders with Papua New Guinea, East Timor, and Malaysia. Other neighbouring countries include Singapore, the Philippines, Australia, and the Indian territory of the Andaman and Nicobar Islands.

The Java Sea is a large (310 000 km^2) shallow sea on the Sunda Shelf. It was formed as sea levels rose at the end of the last Ice Age. The Java Sea lies between the Indonesian islands of Borneo to the north, Java to the south, Sumatra to the west, and Sulawesi to the east. Karimata Strait, in the north-west, links the Java Sea to the South China Sea.

10.6.2   Oceanography

The sea surface temperatures in this region are described by Schweitzer (1993). The sea surface temperatures are around 30 °C in the north and decrease by a few

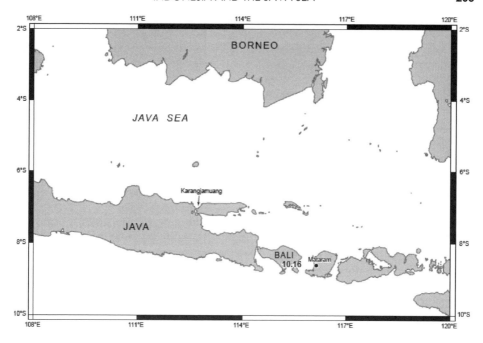

Figure 10.9 Map of Indonesia, Borneo, and the Java Sea. The channel analysed below lies between Bali and Materam.

degrees towards the south. The corresponding sea surface salinities decrease from about 34–35 psu in the surrounding oceans to about 33.5 psu in the central Java Sea (University of Minnesota, 2005).

Egbert and Erofeeva (2002) report the pattern of the M_2 amplitude in the Java Sea showing, in general, values of less than 0.4 m. Hatayama, Awaji, and Akitomo (1996) investigated the characteristics of tides and tidal currents in the area and found that modelled M_2 and K_1 amplitudes compared well with observations.

Relatively strong tidal currents are found in the Java Sea and in the vicinities of the narrow channels such as the Lombok and Malacca Straits (Ray, Egbert, & Erofeeva, 2005). In general, these velocities reflect bathymetry, with shallow water giving rise to rapid currents. Currents in the Banda Sea, the Timor Sea, and part of the Flores Sea tend to rotate anticlockwise; currents elsewhere are generally close to rectilinear, although the deep Pacific currents rotate clockwise. In the east of the region, a significant clockwise residual circulation appears around Buru Island owing primarily to the tidal rectification over variable bottom topography.

10.6.3 Tidal power

The strongest tidal currents listed by the Admiralty occur in the straits between the Java Sea and the Timor Sea. The tides are diurnal (Figure 10.10), and $Fz = 2.25$ has been allocated to the station analysed below.

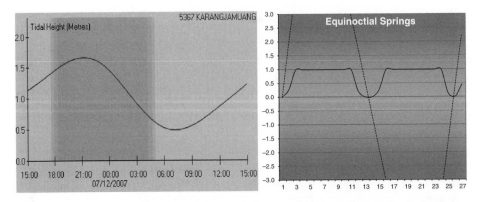

Figure 10.10 (Left) Diurnal tide in the Java Sea (from Admiralty, 2005). (Right) The STEM software results for an Equinoctial tide in the strait between the islands of Bali and Metramor.

Station 10.16 is in the channel that lies between the islands of Bali and Metramor. The data were entered into the STEM software package, which calculated a mean peak Spring tidal current of 2.86 m s^{-1}. The 'standardized' turbine in STEM calculated an annual energy production of $4545 \text{ MW h year}^{-1}$ (Figure 10.10 (right)) at a cost of about £0.033 kW h^{-1}. These results are quite similar to the results from the Gulfs of Kutch and Cambay analysed earlier, and similar comments regarding local construction of the device and distribution will apply.

10.7 East China and Yellow Seas

10.7.1 Geography

The East China Sea (Figure 10.11) is a marginal sea east of mainland China. It is a part of the Pacific Ocean and covers an area of $1\ 249\ 000 \text{ km}^2$. In China, the sea is called the *East Sea*. In South Korea, the sea is sometimes called *South Sea*, but this is more often used to denote only the area near South Korea's southern coast. The East China Sea is bounded on the east by Kyūshū Island and Ryukyu Island, on the south by Taiwan, and on the west by mainland China. It is connected with the South China Sea by the Taiwan Strait and with the Sea of Japan by the Korea Strait; it opens in the north to the Yellow Sea. Territories with borders on the sea (clockwise from north) include South Korea, Japan, Taiwan, and mainland China.

The Yellow Sea is the northern part of the East China Sea. It is located between mainland China and the Korean peninsula. Its name comes from the sand particles that colour its water, originating from the Yellow River. The innermost bay of the Yellow Sea is called the Bohai Sea (previously Pechihli Bay or Chihli Bay). Both the Yellow River (through Shandong Province and its capital, Jinan) and Hai He River (through Beijing and Tianjin) flow into the Yellow Sea.

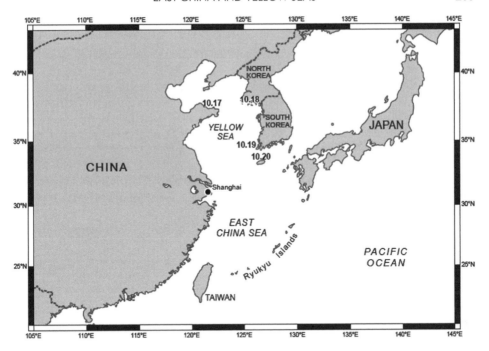

Figure 10.11 The East China and Yellow Seas.

10.7.2 Oceanography

The Tsushima warm current, a branch of the Kuroshio current, flows northwards through the Korea Strait along the Japanese shore. The Liman cold current flows southwards through the Strait of Tartary along the Russian shore. The August sea surface temperatures in this region are described by Schweitzer (1993). The sea surface temperatures are around 30 °C in the south and increase by a few degrees towards the north of the region. The corresponding sea surface salinities have a value of about 33–34 psu in this region (University of Minnesota, 2005).

Then, tidal amplitudes in the East China and Yellow Seas are well described by MSSTATE (2007b). Based, in part, on the data provided by Dietrich *et al.* (1980), it is shown that the diurnal tides exhibit two prominent amphidromes: one at the centre of the northern bay and one in the centre of the main basin of the Yellow Sea. The semi-diurnal tides, however, are much larger and exhibit four amphidromes all located along the western shores of the northern bay and of the Yellow Sea. The tides are greatly amplified along the Korean coast. As a result, the amplitude of the M_2 tide increases from about 0.1–0.5 m along the centre-line to more than 1.00 m on the south-western shores and along the whole of the Korean shores of the East China and Yellow Seas (MSSTATE, 2007b).

MSSTATE (2007c) show that the semi-diurnal tides are dominant in this region. and therefore $Fz = 0.1$ has been applied to all the analyses detailed below.

10.7.3   Tidal power

The tidal currents on the eastern margins of the Yellow Sea are relatively weak and semi-diurnal (Guo & Yanagi, (1998).

Station 10.17 off the headland is listed by the Admiralty with mean peak Spring currents of about $0.97\,\text{m s}^{-1}$. On the coastline of the Korean peninsula, however, very strong tidal currents occur. Stations 10.18 and 10.19 lie off the west coast of South Korea, and the Admiralty list peak mean Spring currents of 1.12 and $1.33\,\text{m s}^{-1}$ respectively. The strongest occur, however, off the south-western coastline of South Korea in channels between the islands. Stations 10.20 and 10.21 show peak mean Spring currents of 4.23 and $4.74\,\text{m s}^{-1}$ respectively, and the STEM software demonstrates that these sites are capable of generating 3949 and $4166\,\text{MW h year}^{-1}$ at rates of £0.038 and $£0.036\,\text{kW h}^{-1}$ respectively.

Goreau *et al.* (2005) report that, in the Uldolmok Strait, South Korea, between Jindo Island and the mainland, tidal currents reach 13 knots. After an initial successful test of two helical turbines in 2002, the Korean government has begun a second phase to produce a megawatt of power. If that is successful, thousands of helical turbines could be used to tap the 3600 megawatts of energy potential in the Strait.

More recently, Lunar Energy (see Chapter 4) have announced an agreement with the utility company Korean Midland Power Co. to develop to install 300 of their 11.5 m diameter RTT turbines in the Wando Hoenggan Waterway on Korea's south-west coastline.

10.8   Bibliography

de Mesquita, A.R. & J. Harari (2007) Propagation of Tides and Circulation of Tidal Currents on the Southeastern Brazilian Shelf. http://www.mares.io.usp.br/aagn/32o1.html [accessed 29 November 2007].

Dietrich, G., K. Kalle, W. krauss & G. Siedler (1980) *General Oceanography* (2nd ed.), John Wiley and Sons, New york, pp. 626.

Egbert, G.D. & Erofeeva, S.Y. (2002) Efficient inverse modeling of barotropic ocean tides. *Journal of Atmospheric and Oceanic Technology* 19:183–204.

Falkland Islands Government (2007) *Geography and Climate.* http://www.bgs.ac.uk/falklands-oil/cultural/geog_clim.htm [accessed 29 November 2007].

Gjevik, B., Nøst, E., & Straume, T. (1994) Model simulations of the tides in the Barents Sea. *J. Geophys. Res.* 99, 3337–3350.

Goreau, T.J., Anderson, S., Gorlov, A., & Kurth, E. (2005) Tidal Energy and Low-head River Power: a Strategy to Use New, Proven Technology to Capture these Vast, Non-polluting Resources for Sustainable Development. Submission to the UN. http://www.globalcoral.org/Tidal%20Energy%20and%20 Low-Head%20River%20Power.htm [accessed 16 April 2007].

Guo, X.G. & Yanagi, T. (1998) Three-dimensional structure of tidal current in the East China Sea and the Yellow Sea. *Journal of Oceanography* 54, 651–668. http://www.terrapub.co.jp/journals/JO/pdf/5406/54060651.pdf [accessed 29 November 2007].

Hatayama, T., Awaji, T., & Akitomo, K. (1996) Tidal currents in the Indonesian Seas and their effect on transport and mixing: Pacific low-latitude western boundary currents and the Indonesian throughflow. *J. Geophys. Res.* 101, 12 353–12 374.

Hollowplanets (2008) *Amplitudes.* http://www.hollowplanets.com/journal/J0007Tides.jpg [accessed 7 July 2008].

Kowalik, Z. (1994) Modeling of topographically amplified diurnal tides in the Nordic Seas. *J. Phys. Oceanogr.* **24**(8), 1717–1731.

Kowalik, Z. & Proshutinsky, A.Yu. (1993) The diurnal tides in the Arctic Ocean. *J. Geophys. Res.* **98**, 16 449–16 468.

Kowalik, Z. & Proshutinsky, A.Yu. (1994) The Arctic Ocean tides, in *The Polar Oceans and their Role in Shaping the Global Environment.* AGU, Washington, DC, pp. 137–158.

Kowalik, Z. & Proshutinsky, A.Yu. (1995) Topographic enhancement of tidal motion in the western Barents Sea. *J. Geophys. Res.* **100**, 2613–2637.

Kowalik, Z., Proshutinsky, A., & Thomas, R.H. (2007) Investigation of the ice-tide interaction in the Arctic Ocean. http://www.ims.uaf.edu/tide/ [accessed 29 November 2007].

MSSTATE (2007a) *Tides in Marginal, Semi-Enclosed and Coastal Seas – Part I: Sea Surface Height. 5.2 North Indian Ocean.* http://www.ssc.erc.msstate.edu/Tides2D/Plots/ion_M2_TOPEX.gif [accessed 23 June 2008].

MSSTATE (2007b) *Tides in Marginal, Semi-Enclosed and Coastal Seas – Part I: Sea Surface Height.* http://www.ssc.erc.msstate.edu/Tides2D/joy_sea.html [accessed 29 November 2007].

MSSTATE (2007c) *Tides in Marginal, Semi-Enclosed and Coastal Seas - Part I: Sea Surface Height. Ratio of the Semi-diurnal to the Diurnal Tides.* http://www.ssc.erc.msstate.edu/Tides2D/Plots/ratio/yes_ratio.gif [accessed 23 June 2008].

Nayak, R.N. & Shetye, S.R. (2003) *Tides in the Gulf of Khambhat, West Coast of India. Estuarine, Coastal and Shelf Science* **57**(1–2), 249–254.

Ray, R.D., Egbert, G.D., and Erofeeva, S.Y. (2005) Tides in the Indonesian Seas. *Oceanography* **18**, 74–78. http://www.tos.org/oceanography/issues/issue_archive/issue_pdfs/18_4/18.4_ray_et_al.pdf [accessed 29 November 2007].

Schweitzer, P.N. (1993) *Modern Average Global Sea-Surface Temperature: US Geological Survey Digital Data Series DDS-10,* US Geological Survey, Reston, VA. <http://geo-nsdi.er.usgs.gov/metadata/digital-data/10/m_augsa.gif> [accessed 18 June 2008].

University of Minnesota (2005) *Sea Water Density and Thermohaline Circulation.* http://www.geo.umn.edu/courses/1006/spring2005/lecturenotes/11%20Density%20and %20Thermohaline.htm [accessed 22 June 2008].

10.9 Appendix

Table 10A.1

Ref	Adm. Ref	Lat	Long	Peak Spring (knots)	Peak Neap (Knots)	FZ_U	M_2 (m s^{-1})	S_2 (m s^{-1})	K_2 (m s^{-1})	Mean P (kW)	Max P (kW)	Output MW h year^{-1}	Cap Fact	Cost £ million	£KW h^{-1}
10.1	SN109D	68 11 N	40 01 E	1.5	0.8	0.00	0.587	0.179	0.029	0	0	0	0	15000	inf
10.2	SN108X	67 40 N	41 05 E	2.8	1.6	0.00	1.122	0.306	0.056	18.67	124.2	163.8	0.019	15000	0.918
10.3	SN108L	66 52 N	41 24 E	2.4	1.3	0.00	0.944	0.281	0.047	7.953	78.02	69.78	0.008	15000	2.155
10.4	SN104D	66 17 N	40 42 E	3.3	1.9	0.00	1.326	0.357	0.066	37.41	203.3	328.2	0.037	15000	0.458
10.5	SN104N	65 53 N	38 51 E	1.3	0.7	0.00	0.51	0.153	0.026	0	0	0	0	15000	inf
10.6	SN255B	00 30 S	47 52 W	4.3	2.0	0.40	1.392	0.508	0.07	73.6	717.7	645.8	0.074	15000	0.233
10.7	SN225D	01 24 S	48 30 W	2.9	1.9	0.40	1.061	0.221	0.053	21.2	221.1	186	0.021	15000	0.809
10.8	SN219B	25 29 S	48 16 W	4.4	0.8	0.00	1.326	0.918	0.066	75.85	468.6	665.5	0.076	15000	0.226
10.9	SN208E	51 21 S	60 39 W	4.1	2.2	0.50	1.339	0.404	0.067	70.93	702.5	622.3	0.071	15000	0.242
10.10	SN207E	51 21 S	58 17 W	1.4	0.7	0.50	0.268	0.089	0.013	5.015	83.2	44	0.005	15000	3.417
10.11	SN403D	02 05 S	41 13 E	1.0	0.4	1.50	0.179	0.077	0.009	0	0	0	0	15000	inf
10.12	SN403F	02 44 S	40 25 E	0.5	0.2	1.50	0.089	0.038	0.004	0	0	0	0	15000	inf
10.13	SN401B	05 43 S	39 14 E	1.1	0.3	1.50	0.327	0.187	0.016	0	0	0	0	15000	inf
10.14	SN433R	22 34 N	69 08 E	5.0	2.9	2.00	0.672	0.179	0.034	432.2	1000	3792	0.433	15000	0.040
10.15	SN435D	21 24 N	72 19 E	5.1	2.8	2.00	0.672	0.196	0.034	439.9	1000	3860	0.441	15000	0.039
10.16	S538	08 45 S	115 45 E	5.6	2.7	2.25	0.529	0.185	0.026	518	1000	4545	0.519	15000	0.033
10.17	SN734B	37 32 N	122 39 E	1.9	1.0	0.10	0.715	0.222	0.036	0.191	42.57	1.68	2E-04	15000	inf
10.18	SN748A	37 33 N	125 43 E	2.2	1.0	0.10	0.789	0.296	0.039	2.736	65.77	24.01	0.003	15000	6.264
10.19	SN752A	34 25 N	125 55 E	2.6	1.1	0.10	0.912	0.37	0.046	10.1	108.3	88.58	0.01	15000	1.697
10.20	SN752C	34 45 N	126 18 E	8.3	4.5	0.10	3.155	0.937	0.158	450.1	1000	3949	0.451	15000	0.038
10.21	SN752B	34 34 N	126 19 E	9.3	4.1	0.10	3.30	1.28	0.17	475	1000	4166	0.48	15000	0.036

Glossary

Abridged and edited from NOAA (2000) *Tide and Current Glossary*. http://tide sandcurrents.noaa.gov/publications/glossary2.pdf.

absolute mean sea level change	An eustatic change in mean sea level relative to a conventional terrestrial coordinate system with the origin at the centre of mass of the Earth.
acoustic Doppler current profiler (ADCP)	A current-measuring instrument employing the transmission of high-frequency acoustic signals in the water. The current is determined by a Doppler shift in the backscatter echo from plankton, suspended sediment, and bubbles, all assumed to be moving with the mean speed of the water. Time gating circuitry is employed, which uses differences in acoustic travel time to divide the water column into range intervals, called bins. The bin determinations allow development of a profile of current speed and direction over most of the water column. The ADCP can be deployed from a moving vessel, tow, buoy, or bottom platform. In the latter configuration, it is non-obtrusive in the water column and thus can be deployed in shipping channels.
age of diurnal inequality	The time interval between the maximum semi-monthly north or south declination of the Moon and the maximum effect of declination upon range of tide or speed of the tidal current. The age may be computed from the harmonic constants by the formula:

$$\text{Age of diurnal inequality} = 0.911(\text{51}° - \text{O1}°) \text{ h}$$

age of Moon	The time elapsed since the preceding new Moon.
age of parallax inequality	The time interval between the perigee of the Moon and the maximum effect of the parallax on range of tide or speed of tidal current. This age may be computed from the harmonic constants by the formula

$$\text{Age of parallax inequality} = 1.837(\text{M2}° - \text{N2}°) \text{ h}$$

age of phase inequality

The time interval between new or full Moon and the maximum effect of these phases on range of tide or speed of tidal current. This age may be computed from the harmonic constants by the formula

$$\text{Age of phase inequality} = 0.984(S2° - M2°)\ h$$

amphidromic point

A point of zero amplitude of the observed or a constituent tide.

amplitude (H)

One-half the range of a constituent tide. By analogy, it may be applied also to the maximum speed of a constituent current.

angular velocity of the Earth's rotation (S)

Time rate of change in angular displacement relative to the fixed stars. It is equal to $0.729211 \times 10^{-4}\ \text{rad s}^{-1}$.

annual inequality

Seasonal variation in water level or current, more or less periodic, owing chiefly to meteorological causes.

anomalistic

Pertaining to the periodic return of the Moon to its perigee or the Earth to its perihelion. The anomalistic month is the average period of the revolution of the Moon around the Earth with respect to lunar perigee, and is approximately 27.554550 days in length. The anomalistic year is the average period of the revolution of the Earth around the Sun with respect to the perihelion, and is approximately 365.2596 days in length.

anomaly

As applied to astronomy, the anomaly is the angle made at any time by the radius vector of a planet or moon with its line of apsides, the angle being reckoned from the perihelion or perigee in the direction of the body's motion. It is called the true anomaly when referred to the actual position of the body, and the mean anomaly when referred to a fictitious body moving with a uniform angular velocity equal to the average velocity of the real body and passing the perihelion or perigee at the same time.

aphelion

The point in the orbit of the Earth (or other planet, etc.) farthest from the Sun.

apogean tides or tidal currents

Tides of decreased range or currents of decreased speed occurring monthly as a result of the Moon being in apogee. The apogean range (An) of the tide is the average range occurring at the time of apogean tides and is most conveniently computed from the harmonic constants. It is smaller than the mean range, where the type of tide is either semi-diurnal or mixed, and is of no practical significance where the type of tide is predominantly diurnal.

apogee

The point in the orbit of the Moon or a man-made satellite farthest from the Earth. The point in the orbit of a satellite farthest from its companion body.

apparent secular trend	The non-periodic tendency of sea level to rise, fall, or remain stationary with time.
apparent time	Time based upon the true position of the Sun, as distinguished from mean time, which is measured by a fictitious Sun moving at a uniform rate. Apparent time is that shown by the sundial, and its noon is the time when the Sun crosses the meridian. The difference between apparent time and mean time is known as the equation of time. Although quite common many years ago, apparent time is seldom used now.
apsides	The points in the orbit of a planet or moon that are the nearest and farthest from the centre of attraction. In the Earth's orbit these are called the perihelion and aphelion, and in the Moon's orbit they are called the perigee and apogee. The line passing through the apsides of an orbit is called the line of apsides.
astronomical time	Time formerly used in astronomical calculations in which the day began at noon rather than midnight. The astronomical day commenced at noon of the civil day of the same date. The hours of the day were numbered consecutively from zero (noon) to 23 (11 a.m. of the following morning). Up to the close of the year 1924, astronomical time was in general use in nautical almanacs. Beginning with the year 1925, the *American Ephemeris* and *Nautical Almanac* and similar publications of other countries abandoned the old astronomical time and adopted Greenwich civil (mean) time for the data given in their tables.
augmenting factor	A factor, used in connection with the harmonic analysis of tides or tidal currents by stencils, to allow for the fact that the tabulated hourly heights or speeds used in the summation for any constituent, other than S, do not in general occur on the exact constituent hours to which they are assigned, but may differ from the same by as much as a half-hour.
automatic tide (water level) gauge	An instrument that automatically registers the rise and fall of the water level. In some instruments, the registration is accomplished by recording the heights at regular time intervals in digital format; in others, by a continuous graph of height against time.
azimuth	The azimuth of a body is the arc of the horizon intercepted between the north or south point and the foot of the vertical circle passing through the body. It is reckoned in degrees from either the north or south point clockwise entirely around the horizon. The azimuth of a current is the direction towards which it is flowing, and is usually reckoned from the north point.

baroclinic   The condition and type of motion when isobaric surfaces of a fluid are not parallel with isopycnal surfaces.

barotropic   The condition and type of motion when isobaric surfaces of a fluid are parallel with isopycnal surfaces.

barycentre   The common centre of mass of the Sun–Earth system or the Moon–Earth system. The distance from the centre of the Sun to the Sun–Earth barycentre is about 280 miles. The distance from the centre of the Earth to the Moon–Earth barycentre is about 2895 miles.

benchmark (BM)   A fixed physical object or mark used as reference for a horizontal or vertical datum. A tidal benchmark is one near a tide station to which the tide staff and tidal datums are referred. A primary benchmark is the principal mark of a group of tidal benchmarks to which the tide staff and tidal datums are referred.

Callippic cycle   A period of four Metonic cycles equal to 76 Julian years, or 27 759 days, devised by Callippus, a Greek astronomer, about 350 BC, as a suggested improvement on the Metonic cycle for a period in which the new and full Moon would recur on the same day of the year. Taking the length of the synodical month as 29.530588 days, there are 940 lunations in the Callippic cycle, with about 0.25 days remaining.

celestial sphere An imaginary sphere of infinite radius, concentric with the Earth, on to which all celestial bodies except the Earth are imagined to be projected.

chart datum   The datum to which soundings on a chart are referred. It is usually taken to correspond to a low water elevation, and its depression below mean sea level is represented by the symbol Z. Since 1980, the chart datum has been implemented to mean lower low water for all marine waters of the United States, its territories, the Commonwealth of Puerto Rico, and the Trust Territory of the Pacific Islands. See datum and National Tidal Datum Convention of 1980.

civil day   A mean solar day commencing at midnight.

civil time   Time in which the day begins at midnight, as distinguished from the former astronomical time in which the day began at noon.

coastline   The low water datum line for purposes of the Submerged Lands Act (Public Law 31).

coastal boundary The mean high water line (MHWL) or mean higher high water line (MHHWL) when tidal lines are used as the coastal boundary. Also, lines used as boundaries inland of and measured from (or points thereon) the MHWL or MHHWL. See marine boundary.

cocurrent line   A line on a map or chart passing through places having the same current hour.

compass direction	Direction as indicated by compass without any correction for compass error. The direction indicated by a compass may differ by a considerable amount from true or magnetic direction.
compass error	The angular difference between a compass direction and the corresponding true direction. The compass error combines the effects of deviation and variation.
component	(1) Same as constituent. (2) That part of a tidal-current velocity that, by resolution into orthogonal vectors, is found to flow in a specified direction.
compound tide	A harmonic tidal (or tidal-current) constituent with a speed equal to the sum or difference of the speeds of two or more elementary constituents. The presence of compound tides is usually attributed to shallow water conditions.
constituent	One of the harmonic elements in a mathematical expression for the tide-producing force and in corresponding formulae for the tide or tidal current. Each constituent represents a periodic change or variation in the relative positions of the Earth, Moon, and Sun. A single constituent is usually written in the form

$$y = A \cos (at + ∀)$$

in which y is a function of time as expressed by the symbol t and is reckoned from a specific origin. The coefficient A is called the amplitude of the constituent and is a measure of its relative importance. The angle $(at + ∀)$ changes uniformly, and its value at any time is called the phase of the constituent. The speed of the constituent is the rate of change in its phase and is represented by the symbol a in the formula. The quantity ∀ is the phase of the constituent at the initial instant from which the time is reckoned. The period of the constituent is the time required for the phase to change through 360° and is the cycle of the astronomical condition represented by the constituent.

constituent day	The time of rotation of the Earth with respect to a fictitious celestial body, representing one of the periodic elements in the tidal forces. It approximates in length the lunar or solar day and corresponds to the period of a diurnal constituent or twice the period of a semi-diurnal constituent. The term is not applicable to the long-period constituents.
constituent hour	One twenty-fourth part of a constituent day.
control current station	A current station at which continuous velocity observations have been made over a minimum period of 29 days. Its purpose is to provide data for computing accepted values of the harmonic and non-harmonic constants essential to tidal-current predictions and circulatory studies. The

data series from this station serves as the control for the reduction of relatively short series from subordinate current stations through the method of comparison of simultaneous observations.

corange line   A line passing through places of equal tidal range.

Coriolis force   A fictional force in the hydrodynamic equations of motion that takes into account the effect of the Earth's rotation on moving objects (including air and water) when viewed with reference to a coordinate system attached to the rotating Earth. The horizontal component is directed 90° to the right (when looking in the direction of motion) in the Northern Hemisphere, and 90° to the left in the Southern Hemisphere. The horizontal component is zero at the equator, and also when the object is at rest relative to the Earth. The Coriolis acceleration $= 2vS$ sin ø, where v is the speed of the object, S is the angular velocity of the Earth, and ø is the latitude. Named for Gaspard Gustave de Coriolis who published his formulation in 1835.

corrected current A relatively short series of current observations from a subordinate station to which a factor is applied to adjust the current to a more representative value based on a relatively long series from a nearby control station. See current and total current.

cotidal hour   The average interval between the Moon's transit over the meridian of Greenwich and the time of the following high water at any place. This interval may be expressed either in solar or lunar time. When expressed in solar time, it is the same as the Greenwich high water interval. When expressed in lunar time, it is equal to the Greenwich high water interval multiplied by the factor 0.966.

cotidal line   A line on a chart or map passing through places having the same tidal hour.

countercurrent   A current usually setting in a direction opposite to that of a main current.

crest   The highest point in a propagating wave.

current   Generally, a horizontal movement of water. Currents may be classified as tidal and non-tidal. Tidal currents are caused by gravitational interactions between the Sun, Moon, and Earth and are part of the same general movement of the sea that is manifested in the vertical rise and fall, called tide. Tidal currents are periodic, with a net velocity of zero over the particular tidal cycle. See tidal wave. Non-tidal currents include the permanent currents in the general circulatory systems of the sea as well as temporary currents arising from more pronounced meteorological variability. Current, however, is also the British equivalent of our non-tidal current.

current constants	Tidal-current relations that remain practically constant for any particular locality. Current constants are classified as harmonic and non-harmonic. The harmonic constants consist of the amplitudes and epochs of the harmonic constituents, and the non-harmonic constants include the velocities and intervals derived directly from the current observations.
current curve	A graphic representation of the flow of the current. In the reversing type of tidal current, the curve is referred to rectangular coordinates with time represented by the abscissa and the speed of the current by the ordinate, the flood speeds being considered as positive and the ebb speeds as negative. In general, the current curve for a reversing tidal current approximates a cosine curve.
current diagram	A graphic table published in the tidal-current tables, showing the speeds of the flood and ebb currents and the times of slack and strengths over a considerable stretch of the channel of a tidal waterway, the times being referred to tide or tidal-current phases at some reference station.
current difference	Difference between the time of slack water (or minimum current) or strength of current in any locality and the time of the corresponding phase of the tidal current at a reference station for which predictions are given in the tidal-current tables.
current ellipse	A graphic representation of a rotary current in which the velocity of the current at different hours of the tidal cycle is represented by radius vectors and vectoral angles. A line joining the extremities of the radius vectors will form a curve roughly approximating an ellipse. The cycle is completed in half a tidal day or in a whole tidal day, according to whether the tidal current is of the semi-diurnal or the diurnal type. A current of the mixed type will give a curve of two unequal loops each tidal day.
current hour	The mean interval between the transit of the Moon over the meridian of Greenwich and the time of strength of flood, modified by the times of slack water (or minimum current) and strength of ebb. In computing the mean current hour, an average is obtained of the intervals for the following phases: flood strength, slack (or minimum) before flood increased by 3.10 h (one-quarter of tidal cycle), slack (or minimum) after flood decreased by 3.10 h, and ebb strength increased or decreased by 6.21 h (one-half of tidal cycle). Before taking the average, the four phases are made comparable by the addition or rejection of such multiples of 12.42 h as may be necessary. The current hour is usually expressed in solar time, but if lunar time is desired, the solar hour should be multiplied by the factor 0.966.

current line   A graduated line attached to a current pole formerly used in measuring the velocity of the current. The line was marked in such a manner that the speed of the current, expressed in knots and tenths, was indicated directly by the length of line carried out by the current pole in a specified interval of time. When marked for a 60 s run, the principal divisions for whole knots were spaced at 101.33 ft, and the subdivisions for tenths of knots were spaced at 10.13 ft. The current line was also known as a log line.

current meter   An instrument for measuring the speed and direction or just the speed of a current. The measurements are Eulerian when the meter is fixed or moored at a specific location. Current meters can be mechanical, electric, electromagnetic, acoustic, or any combination thereof.

current pole   A pole used in observing the velocity of the current.

current station   The geographic location at which current observations are conducted. Also, the facilities used to make current observations. These may include a buoy, ground tackle, current meters, recording mechanism, and radio transmitter. See control current station and subordinate current station (1).

cyclonic ring   A meander breaking off from the main oceanic current and spinning in a counterclockwise direction in the Northern Hemisphere (clockwise in the Southern Hemisphere).

datum (vertical)   For marine applications, a base elevation used as a reference from which to reckon heights or depths. It is called a tidal datum when defined in terms of a certain phase of the tide. Tidal datums are local datums and should not be extended into areas that have differing hydrographic characteristics without substantiating measurements. In order that they may be recovered when needed, such datums are referenced to fixed points known as benchmarks. See chart datum.

datum of tabulation   A permanent base elevation at a tide station to which all water level measurements are referred. The datum is unique to each station and is established at a lower elevation than the water is ever expected to reach. It is referenced to the primary benchmark at the station and is held constant regardless of changes to the water level gauge or tide staff. The datum of tabulation is most often at the zero of the first tide staff installed.

day   The period of rotation of the Earth. There are several kinds of day, depending on whether the Sun, Moon, or other object or location is used as the reference for the rotation. See constituent day, lunar day, sidereal day, and solar day.

daylight saving time   A time used during the summer months, in some localities, in which clocks are advanced 1 h from the usual standard time.

decibar	The practical unit for pressure in the ocean, equal to 10 cbar, and the approximate pressure produced by each metre of overlying water.
declination	Angular distance north or south of the celestial equator, taken as positive when north of the equator and negative when south. The Sun passes through its declinational cycle once a year, reaching its maximum north declination of approximately 23½° about 21 June and its maximum south declination of approximately 23½° about 21 December. The Moon has an average declinational cycle of 27½ days, which is called a tropical month. Tides or tidal currents occurring near the times of maximum north or south declination of the Moon are called tropic tides or tropic currents, and those occurring when the Moon is over the equator are called equatorial tides or equatorial currents. The maximum declination reached by the Moon in successive months depends upon the longitude of the Moon's node, and varies from 28½° when the longitude of the ascending node is 0° to 18½° when the longitude of the node is 180°. The node cycle, or the time required for the node to complete a circuit of 360° of longitude, is approximately 18.6 years.
declinational reduction	A processing of observed high and low waters or flood and ebb tidal currents to obtain quantities depending upon changes in the declination of the Moon, such as tropic ranges or speeds, height or speed inequalities, and tropic intervals.
density, *in situ* ($D_{s,t,p}$)	Mass per unit volume. The reciprocal of specific volume. In oceanography, the density of sea water, when expressed in g cm^{-3}, is numerically equivalent to specific gravity and is a function of salinity, temperature, and pressure. See specific volume anomaly, thermosteric anomaly, sigma-*t*, and sigma-zero.
deviation (of compass)	The deflection of the needle of a magnetic compass owing to masses of magnetic metal within a ship on which the compass is located. This deflection varies with different headings of the ship. The deviation is called easterly and marked plus if the deflection is to the right of magnetic north, and is called westerly and marked minus if it is to the left of magnetic north. A deviation table is a tabular arrangement showing the amount of deviation for different headings of the ship.
direct method	A tidal datum computation method. Datums are determined directly by comparison with an appropriate control for the available part of the tidal cycle. It is usually used only when a full range of tidal values are not available; for example, direct mean high water, when low waters are not recorded.
direction of current	Same as set.
direction of wind	Direction from which the wind is blowing.

diurnal	Having a period or cycle of approximately one tidal day. Thus, the tide is said to be diurnal when only one high water and one low water occur during a tidal day, and the tidal current is said to be diurnal when there is a single flood and a single ebb period of a reversing current in the tidal day. A rotary current is diurnal if it changes its direction through all points of the compass once each tidal day. A diurnal constituent is one that has a single period in the constituent day. The symbol for such a constituent is the subscript 1. See stationary wave theory and type of tide.
diurnal inequality	The difference in height of the two high waters or of the two low waters of each tidal day; also, the difference in speed between the two flood tidal currents or the two ebb currents of each tidal day. The difference changes with the declination of the Moon, and, to a lesser extent, with the declination of the Sun. In general, the inequality tends to increase with increasing declination, either north or south, and to diminish as the Moon approaches the equator. Mean diurnal high water inequality (DHQ) is one-half the average difference between the two high waters of each tidal day observed over the national tidal datum epoch. It is obtained by subtracting the mean of all the high waters from the mean of the higher high waters. Mean diurnal low water inequality (DLQ) is one-half the average difference between the two low waters of each tidal day observed over the national tidal datum epoch. It is obtained by subtracting the mean of the lower low waters from the mean of all the low waters. Tropic high water inequality (HWQ) is the average difference between the two high waters of each tidal day at the times of tropic tides. Tropic low water inequality (LWQ) is the average difference between the two low waters of each tidal day at the times of tropic tides. Mean and tropic inequalities, as defined above, are applicable only when the type of tide is either semi-diurnal or mixed. Diurnal inequality is sometimes called declinational inequality.
diurnal tide level	A tidal datum midway between mean higher high water and mean lower low water.
double ebb	An ebb tidal current having two maxima of speed separated by a smaller ebb speed.
double flood	A flood tidal current having two maxima of speed separated by a smaller flood speed.
double tide	A double-headed tide, that is, a high water consisting of two maxima of nearly the same height separated by a relatively small depression, or a low water consisting of two minima separated by a relatively small elevation.

duration of flood and duration of ebb	Duration of flood is the interval of time in which a tidal current is flooding, and duration of ebb is the interval in which it is ebbing, these intervals being reckoned from the middle of the intervening slack waters or minimum currents. Together they cover, on average, a period of 12.42 h for a semi-diurnal tidal current or a period of 24.84 h for a diurnal current. In a normal semi-diurnal tidal current, the duration of flood and duration of ebb will each be approximately equal to 6.21 h, but the times may be modified greatly by the presence of non-tidal flow. In a river, the duration of ebb is usually longer than the duration of flood because of freshwater discharge, especially during spring months when snow and ice melt are predominant influences.
duration of rise and duration of fall	Duration of rise is the interval from low water to high water, and duration of fall is the interval from high water to low water. Together they cover, on average, a period of 12.42 h for a semi-diurnal tide or a period of 24.84 h for a diurnal tide. In a normal semi-diurnal tide, duration of rise and duration of fall will each be approximately equal to 6.21 h, but in shallow waters and in rivers there is a tendency for a decrease in duration of rise and a corresponding increase in duration of fall.
dynamic decimetre	See geopotential as preferred term.
dynamic depth (height)	See geopotential difference as preferred term.
earth tide	Periodic movement of the Earth's crust caused by gravitational interactions between the Sun, Moon, and Earth.
ebb axis	Average set of the current at ebb strength.
ebb current (ebb)	The movement of a tidal current away from shore or down a tidal river or estuary. In the mixed type of reversing tidal current, the terms greater ebb and lesser ebb are applied respectively to ebb tidal currents of greater and lesser speed each day. The terms maximum ebb and minimum ebb are applied to the maximum and minimum speeds of a current running continuously ebb, the speed alternately increasing and decreasing without coming to a slack or reversing. The expression maximum ebb is also applicable to any ebb current at the time of greatest speed. See ebb strength.
ebb interval	The interval between the transit of the Moon over the meridian of a place and the time of the following ebb strength.
ebb strength (strength of ebb)	Phase of the ebb tidal current at the time of maximum speed. Also, the speed at this time. See strength of current.
eccentricity of orbit	Ratio of the distance from the centre to the focus of an elliptical orbit to the length of the semi-major axis. The eccentricity of orbit $= \%1 - (B/A)\,2$, where A and B are respectively the semi-major and semi-minor axes of the orbit.

ecliptic   The intersection of the plane of the Earth's orbit with the celestial sphere.

eddy   A quasi-circular movement of water whose area is relatively small in comparison with the current with which it is associated.

edge waves   Waves moving between zones of high and low breakers along the shoreline. Edge waves contribute to changes in water level along the shoreface, which helps to control the spacing of rip currents. See long-shore current and rip current.

Ekman spiral   A logarithmic spiral (when projected on a horizontal plane) formed by the heads of current velocity vectors at increasing depths. The current vectors become progressively smaller with depth. They spiral to the right (looking in the direction of flow) in the Northern Hemisphere and to the left in the Southern Hemisphere with increasing depth. Theoretically, in deep water, the surface current vector sets 45° and the total mass transport sets 90° from the direction towards which the wind is blowing. Flow opposite to the surface current occurs at the so-called 'depth of frictional resistance'. The phenomenon occurs in wind drift currents in which only the Coriolis and frictional forces are significant. Named for Vagn Walfrid Ekman, who, assuming a constant eddy viscosity, steady wind stress, and unlimited water depth and extent, derived the effect in 1905.

elimination   One of the final processes in the harmonic analysis of tides in which preliminary values for the harmonic constants of a number of constituents are cleared of the residual effects of each other.

epoch   (1) Also known as phase lag. Angular retardation of the maximum of a constituent of the observed tide (or tidal current) behind the corresponding maximum of the same constituent of the theoretical equilibrium tide. It may also be defined as the phase difference between a tidal constituent and its equilibrium argument. As referred to the local equilibrium argument, its symbol is 6. When referred to the corresponding Greenwich equilibrium argument, it is called the Greenwich epoch and is represented by G. A Greenwich epoch that has been modified to adjust to a particular time meridian for convenience in the prediction of tides is represented by g or by $6N$. The relations between these epochs may be expressed by the following formulae:

$$G = 6 + pL$$
$$g = 6N = G - aS/15$$

in which L is the longitude of the place and S is the longitude of the time meridian, these being taken as positive for west longitude and negative for east longitude; p is the number of constituent periods in the constituent day and is equal to 0 for all long-period constituents, 1 for diurnal constituents,

2 for semi-diurnal constituents, and so forth; and a is the hourly speed of the constituent, all angular measurements being expressed in degrees. (2) As used in tidal datum determination, it is a 19 year cycle over which tidal height observations are meaned in order to establish the various datums. As there are periodic and apparent secular trends in sea level, a specific 19 year cycle (the national tidal datum epoch) is selected so that all tidal datum determinations throughout the United States, its territories, the Commonwealth of Puerto Rico, and the Trust Territory of the Pacific Islands will have a common reference.

equation of time	Difference between mean and apparent time. From the beginning of the year until near the middle of April, mean time is ahead of apparent time, the difference reaching a maximum of about 15 min near the middle of February. From the middle of April to the middle of June, mean time is behind apparent time, but the difference is less than 5 min. From the middle of June to the first part of September, mean time is again ahead of apparent time, with a maximum difference of less than 7 min. From the 1 September until the later part of December, mean time is again behind apparent time, the difference reaching a maximum of nearly 17 min in the early part of November. The equation of time for each day in the year is given in the *American Ephemeris* and *Nautical Almanac*.
equatorial tidal currents	Tidal currents occurring semi-monthly as a result of the Moon being over the equator. At these times, the tendency of the Moon to produce a diurnal inequality in the tidal current is at a minimum.
equatorial tides	Tides occurring semi-monthly as a result of the Moon being over the equator. At these times, the tendency of the Moon to produce a diurnal inequality in the tide is at a minimum.
equilibrium argument	The theoretical phase of a constituent of the equilibrium tide. It is usually represented by the expression $(V + u)$, in which V is a uniformly changing angular quantity involving multiples of the hour angle of the mean sun, the mean longitudes of the Moon and Sun, and the mean longitude of lunar or solar perigee, and u is a slowly changing angle depending upon the longitude of the Moon's node. When pertaining to an initial instant of time, such as the beginning of a series of observations, it is expressed by $(V_0 + u)$.
equilibrium theory	A model under which it is assumed that the waters covering the face of the Earth instantly respond to the tide-producing forces of the Moon and Sun to form a surface of equilibrium under the action of these forces. The model disregards friction, inertia, and the irregular distribution of the landmasses of the Earth. The theoretical tide formed under these conditions is known as the equilibrium tide.

equilibrium tide   Hypothetical tide due to the tide-producing forces under the equilibrium theory.

equinoctial   The celestial equator.

Equinoctial tides   Tides occurring near the times of the equinoxes.

equinoxes   The two points in the celestial sphere where the celestial equator intersects the ecliptic; also, the times when the Sun crosses the equator at these points. The vernal equinox is the point where the Sun crosses the equator from south to north and occurs about 21 March. Celestial longitude is reckoned eastward from the vernal equinox. The autumnal equinox is the point where the Sun crosses the equator from north to south and occurs about 23 September.

establishment of the port   Also known as high water, full and change (HWF&C). Average high water interval on days of the new and full Moon. This interval is also sometimes called the common or vulgar establishment to distinguish it from the corrected establishment, the latter being the mean of all the high water intervals. The latter is usually 10–15 min less than the common establishment.

estuary   An embayment of the coast in which fresh river water entering at its head mixes with the relatively saline ocean water. When tidal action is the dominant mixing agent, it is usually termed a tidal estuary. Also, the lower reaches and mouth of a river emptying directly into the sea where tidal mixing takes place. The latter is sometimes called a river estuary.

Eulerian measurement   Observation of a current with a device fixed relative to the flow.

eustatic sea level rate   The worldwide change in sea level elevation with time. The changes are due to such causes as glacial melting or formation, thermal expansion or contraction of sea water, etc.

evection   A perturbation of the Moon, depending upon the alternate increase and decrease in the eccentricity of its orbit, which is always a maximum when the Sun is passing the Moon's line of apsides and a minimum when the Sun is at right angles to it. The principal constituents in the tide resulting from the evectional inequality are <2, 82, and D_1.

first reduction   A method of determining high and low water heights, time intervals, and ranges from an arithmetic mean without adjustment to a long-term series through comparison of simultaneous observations.

float well   A stilling well in which the float of a float-actuated water level gauge operates. See stilling well.

flood axis   The average set of the tidal current at strength of flood.

flood current (flood)   The movement of a tidal current towards the shore or up a tidal river or estuary. In the mixed type of reversing current, the terms greater flood and lesser flood are applied respectively to the two flood currents of greater and lesser speed of each day. The expression maximum flood is applicable to any flood current at the time of greatest speed. See flood strength.

flood interval	The interval between the transit of the Moon over the meridian of a place and the time of the following flood strength.
flood strength (strength of flood)	Phase of the flood tidal current at the time of maximum speed. Also, the speed at this time. See strength of current.
flow	The British equivalent of the United States total current. Flow is the combination of tidal stream and current.
flushing time	The time required to remove or reduce (to a permissible concentration) any dissolved or suspended contaminant in an estuary or harbour.
forced wave	A wave generated and maintained by a continuous force.
fortnight	The time elapsed between the new and full Moons. Half a synodical month, or 14.765294 days. See synodical month.
Fourier series	A series proposed by the French mathematician Fourier about the year 1807. The series involves the sines and cosines of whole multiples of a varying angle and is usually written in the following form:

$$y = AB + A1 \sin x + A2 \sin 2x + A3 \sin 3x$$
$$+ \cdots + B1 \cos x + B2 \cos 2x + B3 \cos 3x + \cdots$$

	By taking a sufficient number of terms, the series may be assumed to represent any periodic function of x.
free wave	A wave that continues to exist after the generating force has ceased to act.
gas purged pressure gauge	A type of water level gauge in which gas, usually nitrogen, is emitted from a submerged orifice at a constant rate. Fluctuations in hydrostatic pressure due to changes in water level modify the recorded emission rate. Same as bubbler tide (water level) gauge.
geodetic datum	See national geodetic vertical datum of 1929 (NGVD 1929) and North American vertical datum of 1988 (NAVD 1988).
geopotential	The unit of geopotential difference, equal to the gravity potential of $1 \, \text{m}^2 \, \text{s}^{-2}$, or $1 \, \text{J kg}^{-1}$.
geopotential anomaly (D)	The excess in geopotential difference over the standard geopotential difference (at a standard specific volume at 35 parts per thousand (‰) and $0\,°\text{C}$) between isobaric surfaces.
geopotential difference	The work per unit mass gained or required in moving a unit mass vertically from one geopotential surface to another. See geopotential, geopotential anomaly, and geopotential topography.
geopotential (equipotential) surface	A surface that is everywhere normal to the acceleration of gravity.
geopotential topography	The topography of an equiscalar (usually isobaric) surface in terms of geopotential difference. As depicted on maps, isopleths are formed by the intersection of the isobaric surface with a series of geopotential surfaces. Thus, the field of isopleths represents variations in the geopotential anomaly of the isobaric surface above a chosen reference isobaric surface (such as a level of no motion).

geostrophic flow	A solution of the relative hydrodynamic equations of motion in which it is assumed that the horizontal component of the Coriolis force is balanced by the horizontal component of the pressure gradient force.
gradient flow	A solution of the relative hydrodynamic equations of motion in which only the horizontal Coriolis, pressure gradient, and centrifugal forces are considered.
gravity wave	A wave for which the restoring force is gravity.
great diurnal range (GT)	The difference in height between mean higher high water and mean lower low water. The expression may also be used in its contracted form, diurnal range.
great tropic range (GC)	The difference in height between tropic higher high water and tropic lower low water. The expression may also be used in its contracted form, tropic range.
Greenwich argument	Equilibrium argument computed for the meridian of Greenwich.
Greenwich interval	An interval referred to the transit of the Moon over the meridian of Greenwich, as distinguished from the local interval which is referred to the Moon's transit over the local meridian. The relation in hours between Greenwich and local intervals may be expressed by the formula

$$\text{Greenwich interval} = \text{local interval} + 0.069L$$

where L is the west longitude of the local meridian in degrees. For east longitude, L is to be considered negative.

Gregorian calendar	The modern calendar in which every year divisible by 4 (excepting century years) and every century year divisible by 400 are bissextile (or leap) years with 366 days. All other years are common years with 365 days. The average length of this year is therefore 365.2425 days, which agrees very closely with the length of the tropical year (the period of changes in seasons). The Gregorian calendar was introduced by Pope Gregory in 1582, and immediately adopted by the Catholic countries in place of the Julian calendar previously in use. In making the change, it was ordered that the day following 4 October 4 1582 of the Julian calendar be designated 15 October 1582 of the Gregorian calendar, the 10 days being dropped in order that the vernal equinox would fall on 21 March. The Gregorian calendar was not adopted by England until 1752, but is now in general use throughout the world.
Gulf coast low water datum (GCLWD)	A tidal datum. Used as chart datum from 14 November 1977 to 27 November 1980 for the coastal waters of the Gulf coast of the United States. GCLWD is defined as mean lower low water when the type of tide is mixed and mean low water (now mean lower low water) when the type of tide is diurnal. See National Tidal Datum Convention of 1980.
h	Rate of change (as of 1 January 1900) in mean longitude of the Sun. $h = 0.04106864°$ per solar hour.

harmonic analysis	The mathematical process by which the observed tide or tidal current at any place is separated into basic harmonic constituents.
harmonic analyser	A machine designed for the resolution of a periodic curve into its harmonic constituents. Now performed by electronic digital computer.
harmonic constants	The amplitudes and epochs of the harmonic constituents of the tide or tidal current at any place.
harmonic function	In its simplest form, a quantity that varies as the cosine of an angle that increases uniformly with time. It may be expressed by the formula

$$y = A \cos at$$

in which y is a function of time (t), A is a constant coefficient, and a is the rate of change in the angle at.

harmonic prediction	Method of predicting tides and tidal currents by combining the harmonic constituents into a single tide curve. The work is usually performed by electronic digital computer.
head	The difference in water level at either end of a strait, channel, inlet, etc.
head of tide	The inland or upstream limit of water affected by the tide.
high water (HW)	The maximum height reached by a rising tide. The high water is due to the periodic tidal forces and the effects of meteorological, hydrologic, and/or oceanographic conditions. For tidal datum computational purposes, the maximum height is not considered a high water unless it contains a tidal high water.
high water line	The intersection of the land with the water surface at an elevation of high water.
higher high water (HHW)	The highest of the high waters (or single high water) of any specified tidal day owing to the declinational effects of the Moon and Sun.
higher low water (HLW)	The highest of the low waters of any specified tidal day owing to the declinational effects of the Moon and Sun.
hydraulic current	A current in a channel caused by a difference in the surface elevation at the two ends. Such a current may be expected in a strait connecting two bodies of water in which the tides differ in time or range. The current in the East River, New York, connecting Long Island Sound and New York Harbour, is an example.
hydrographic datum	A datum used for referencing depths of water and the heights of predicted tides or water level observations. Same as chart datum. See datum.
Indian Spring low water	A datum originated by Professor G.H. Darwin when investigating the tides of India. It is an elevation depressed below mean sea level by an amount equal to the sum of the amplitudes of the harmonic constituents M_2, S_2, K_1, and O_1.
inequality	A systematic departure from the mean value of a tidal quantity. See diurnal inequality, parallax inequality, and phase inequality.

inertial flow   A solution of the relative hydrodynamic equations of motion in which only the horizontal component of the Coriolis and centrifugal forces are balanced. This anticyclonic flow results from a sudden application and release of a driving force, which then allows the system to continue on under its own momentum without further interference. The period of rotation is $2B/2S$ sin ø, where $S = 0.729211 \times 10^{-4}$ rad s^{-1}, and ø = latitude.

internal tide   A tidal wave propagating along a sharp density discontinuity, such as a thermocline, or in an area of gradually (vertically) changing density.

International Hydrographic Organization (formerly Bureau)   An institution consisting of representatives of a number of nations organized for the purpose of coordinating the hydrographic work of the participating governments. It had its origin in the International Hydrographic Conference in London in 1919. It has permanent headquarters in the Principality of Monaco and is supported by funds provided by the member nations. Its principal publications include the *Hydrographic Review* and special publications on technical subjects.

intertidal zone   (Technical definition) The zone between the mean higher high water and mean lower low water lines.

interval   See lunitidal interval and lunicurrent interval.

inverse barometer effect   The inverse response of sea level to changes in atmospheric pressure. A static reduction of 1.005 mbar in atmospheric pressure will cause a stationary rise of 1 cm in sea level.

isanostere   An isopleth of either specific volume anomaly or thermosteric anomaly.

isobar   An isopleth of pressure.

isobaric surface   A surface of constant or uniform pressure.

isohaline   An isopleth of salinity. Constant or uniform in salinity.

isopleth   A line of constant or uniform value of a given quantity. See isanostere, isobar, isohaline, isopycnic, and isotherm.

isopycnic   An isopleth of density. Constant or uniform in density.

isotherm   An isopleth of temperature.

J_1   Smaller lunar elliptic diurnal constituent. This constituent, with M_1, modulates the amplitudes of the declinational K_1 for the effect of the Moon's elliptical orbit:

$$\text{Speed} = T + s + h - p = 15.5854433° \text{ per solar hour}$$

Julian calendar   A calendar introduced by Julius Caesar in the year 45 BC, and slightly modified by Augustus a few years later. This calendar provided that the common year should consist of 365 days and that every fourth year, now known as a bissextile or leap year, should contain 366 days, making the average length of the year 365.25 days. It differs from the modern or Gregorian calendar in having every fourth year a leap year, while, in the modern calendar, century years not divisible by 400 are common years. See Gregorian calendar.

Julian date	Technique for the identification of successive days of the year when monthly notation is not desired. This is especially applicable in computer data processing and acquisition where indexing is necessary.
K_1	Lunisolar diurnal constituent. This constituent, with O_1, expresses the effect of the Moon's declination. They account for diurnal inequality and, at extremes, diurnal tides. With P_1, it expresses the effect of the Sun's declination:

$$\text{Speed} = T + h = 15.0410686° \text{ per solar hour}$$

K_2	Lunisolar semi-diurnal constituent. This constituent modulates the amplitude and frequency of M_2 and S_2 for the declinational effect of the Moon and Sun respectively:

$$\text{Speed} = 2T + 2h = 30.0821373° \text{ per solar hour}$$

kappa	Name of Greek letter used as the symbol for a constituent phase lag or epoch when referred to the local equilibrium argument and frequently taken to mean the same as local epoch.
knot	A speed unit of 1 international nautical mile (1852.0 m or 6076.11549 international ft) per hour.
L_2	Smaller lunar elliptic semi-diurnal constituent. This constituent, with N_2, modulates the amplitude and frequency of M_2 for the effect of variation in the Moon's orbital speed owing to its elliptical orbit:

$$\text{Speed} = 2T - s + 2h - p$$
$$= 29.5284789¼ \text{ per solar hour}$$

lagging of tide	The periodic retardation in the time of occurrence of high and low water owing to changes in the relative positions of the Moon and Sun.
Lagrangian measurement	Observation of a current with a device moving with the current.
lambda	Smaller lunar evectional constituent. This constituent, with <2, :2, and (S_2), modulates the amplitude and frequency of M_2 for the effects of variation in solar attraction of the Moon. This attraction results in a slight pear-shaped lunar ellipse and a difference in lunar orbital speed between motion towards and away from the Sun:

$$\text{Speed} = 2T - s + p = 29.4556253° \text{ per solar hour}$$

latitude	The angular distance between a terrestrial position and the equator measured northward or southward from the equator along a meridian of longitude.
leap year	A calendar year containing 366 days. According to the present Gregorian calendar, all years with the date number divisible by 4 are leap years, except century years. The latter are leap years when the date number is divisible by 400.

level of no motion	A level (or layer) at which it is assumed that an isobaric surface coincides with a geopotential surface. A level (or layer) at which there is no horizontal pressure gradient force.
local time	Time in which noon is defined by the transit of the Sun over the local meridian, as distinguished from standard time which is based upon the transit of the Sun over a standard meridian. Local time may be either mean or apparent, according to whether reference is to the mean or actual Sun. Local time was in general use in the United States until 1883, when standard time was adopted. The use of local time in other parts of the world has also been practically abandoned in favour of the more convenient standard time.
log line	A graduated line used to measure the speed of a vessel through the water or to measure the velocity of the current from a vessel at anchor. See current line.
long-period constituent	A tidal or tidal-current constituent with a period that is independent of the rotation of the Earth but that depends upon the orbital movement of the Moon or the Earth. The principal lunar long-period constituents have periods approximating a month and half a month, and the principal solar long-period constituents have periods approximating a year and half a year.
long-period waves (long waves)	Forced or free waves whose lengths are much longer than the water depth. See tidal wave and tsunami.
longitude	Angular distance along a great circle of reference reckoned from an accepted origin to the projection of any point on that circle. Longitude on the Earth's surface is measured on the equator east and west of the meridian of Greenwich and may be expressed either in degrees or in hours, the hour being taken as the equivalent of 15° of longitude. Celestial longitude is measured in the ecliptic eastward from the vernal equinox. The mean longitude of a celestial body moving in an orbit is the longitude that would be attained by a point moving uniformly in the circle of reference angular velocity to that of the body, with the initial position of the point so taken that its longitude would be the same as that of the body at a certain specified position in its orbit. With a common initial point, the mean longitude of a body will be the same in whatever circle it may be reckoned.
low water (LW)	The minimum height reached by a falling tide. The low water is due to the periodic tidal forces and the effects of meteorological, hydrologic, and/or oceanographic conditions. For tidal datum computational purposes, the minimum height is not considered a low water unless it contains a tidal low water.
low water Equinoctial Springs	Low water Springs near the times of the equinoxes. Expressed in terms of the harmonic constants, it is an elevation depressed below mean sea level by an amount equal to the sum of the amplitudes of the constituents M_2, S_2, and K_2.

lower high water (LHW)	The lowest of the high waters of any specified tidal day owing to the declinational effects of the Moon and Sun.
lower low water (LLW)	The lowest of the low waters (or single low water) of any specified tidal day owing to the declinational effects of the Moon and Sun.
lowest astronomical tide	As defined by the International Hydrographic Organization, the lowest tide level that can be predicted to occur under average meteorological conditions and under any combination of astronomical conditions.
lunar cycle	An ambiguous expression that has been applied to various cycles associated with the Moon's motion. See Callippic cycle, Metonic cycle, node cycle, and synodical month.
lunar day	The time of the rotation of the Earth with respect to the Moon, or the interval between two successive upper transits of the Moon over the meridian of a place. The mean lunar day is approximately 24.84 solar hours in length, or 1.035 times as great as the mean solar day.
lunar interval	The difference in time between the transit of the Moon over the meridian of Greenwich and a local meridian. The average value of this interval, expressed in hours, is $0.069L$, where L is the local longitude in degrees, positive for west longitude and negative for east. The lunar interval equals the difference between the local and Greenwich interval of a tide or current phase.
lunar nodes	The points where the plane of the Moon's orbit intersects the ecliptic. The point where the Moon crosses in going from south to north is called the ascending node, and the point where the crossing is from north to south is called the descending node. References are usually made to the ascending node, which, for brevity, may be called the node.
lunar tide	That part of the tide on the Earth that is due solely to the Moon, as distinguished from that part due to the Sun.
lunar time	Time based upon the rotation of the Earth relative to the Moon. See lunar day.
lunicurrent interval	The interval between the Moon's transit (upper or lower) over the local or Greenwich meridian and a specified phase of the tidal current following the transit. Examples are strength of flood interval and strength of ebb interval, which may be abbreviated to flood interval and ebb interval respectively. The interval is described as local or Greenwich according to whether the reference is to the Moon's transit over the local or Greenwich meridian. When not otherwise specified, the reference is assumed to be local.
lunisolar tides	Harmonic tidal constituents K_1 and K_2, which are derived partly from the development of the lunar tide and partly from the solar tide, the constituent speeds being the same in both cases. Also, the lunisolar synodic fortnightly constituent MS_f.

lunitidal   The interval between the Moon's transit (upper or lower) over the
interval   local or Greenwich meridian and the following high or low water.
The average of all high water intervals for all phases of the Moon
is known as the mean high water lunitidal interval and is abbreviated to high water interval (HWI). Similarly, the mean low water
lunitidal interval is abbreviated to low water interval (LWI). The
interval is described as local or Greenwich according to whether
the reference is to the transit over the local or Greenwich meridian.
When not otherwise specified, the reference is assumed to be local.
When there is considerable diurnal inequality in the tide, separate
intervals may be obtained for the higher high waters, lower high
waters, higher low waters, and lower low waters. These are designated respectively as the higher high water interval (HHWI), lower
high water interval (LHWI), higher low water interval (HLWI), and
lower low water interval (LLWI). In such cases, and also when the
tide is diurnal, it is necessary to distinguish between the upper and
lower transit of the Moon with reference to its declination. Intervals referred to the Moon's upper transit at the time of its north
declination or the lower transit at the time of south declination are
marked *a*. Intervals referred to the Moon's lower transit at the time
of its north declination or to the upper transit at the time of south
declination are marked *b*.

M_1 Smaller lunar elliptic diurnal constituent. This constituent, with
J_1, modulates the amplitude of the declinational K_1 for the effect
of the Moon's elliptical orbit. A slightly slower constituent, designated (M_1), with Q_1, modulates the amplitude and frequency of the
declinational O_1 for the same effect:

$$\text{Speed} = T - s + h + p = 14.4966939° \text{ per solar hour}$$

M_2 Principal lunar semi-diurnal constituent. This constituent represents
the rotation of the Earth with respect to the Moon:

$$\text{Speed} = 2T - 2s + 2h = 28.9841042° \text{ per solar hour}$$

M_3 Lunar terdiurnal constituent. A shallow water compound constituent. See shallow water constituent:

$$\text{Speed} = 3T - 3s + 3h = 43.4761563° \text{ per solar hour}$$

M_4, M_6, M_8 Shallow water overtides of the principal lunar constituent. See
shallow water constituent:

$$\text{Speed of } M_4 = 2M_2 = 4T - 4s + 4h$$
$$= 57.9682084¼ \text{ per solar hour}$$
$$\textit{Speed of } M_6 = 3M_2 = 6T - 6s + 6h$$
$$= 86.952312.7¼ \text{ per solar hour}$$
$$\textit{Speed of } M_8 = 4M_2 = 8T - 8s + 8h$$
$$= 115.9364169¼ \text{ per solar hour}$$

magnetic azimuth	Azimuth reckoned from the magnetic north or magnetic south. See magnetic direction.
magnetic direction	Direction as indicated by a magnetic compass after correction for deviation but without correction for variation.
marigram	A graphic record of the rise and fall of water level. The record is in the form of a curve in which time is generally represented on the abscissa and the height of the water level on the ordinate. See tide curve.
mean current hour	Same as current hour.
mean diurnal tide level (MDTL)	A tidal datum. The arithmetic mean of mean higher high water and mean lower low water.
mean high water (MHW)	A tidal datum. The average of all the high water heights observed over the national tidal datum epoch. For stations with shorter series, comparison of simultaneous observations with a control tide station is made in order to derive the equivalent datum of the national tidal datum epoch.
mean high water line (MHWL)	The line on a chart or map that represents the intersection of the land with the water surface at the elevation of mean high water. See shoreline.
mean higher high water (MHHW)	A tidal datum. The average of the higher high water height of each tidal day observed over the national tidal datum epoch. For stations with shorter series, comparison of simultaneous observations with a control tide station is made in order to derive the equivalent datum of the national tidal datum epoch.
mean higher high water line (MHHWL)	The line on a chart or map that represents the intersection of the land with the water surface at the elevation of mean higher high water.
mean low water (MLW)	A tidal datum. The average of all the low water heights observed over the national tidal datum epoch. For stations with shorter series, comparison of simultaneous observations with a control tide station is made in order to derive the equivalent datum of the national tidal datum epoch.
mean low water line (MLWL)	The line on a chart or map that represents the intersection of the land with the water surface at the elevation of mean low water.
mean low water Springs (MLWS)	A tidal datum. Frequently abbreviated Spring low water. The arithmetic mean of the low water heights occurring at the time of Spring tides observed over the national tidal datum epoch. It is usually derived by taking an elevation depressed below the half-tide level by an amount equal to one-half the Spring range of tide, necessary corrections being applied to reduce the result to a mean value. This datum is used, to a considerable extent, for hydrographic work outside the United States and is the level of reference for the Pacific approaches to the Panama Canal.

mean lower low water (MLLW) A tidal datum. The average of the lower low water height of each tidal day observed over the national tidal datum epoch. For stations with shorter series, comparison of simultaneous observations with a control tide station is made in order to derive the equivalent datum of the national tidal datum epoch.

mean lower low water line (MLLWL) The line on a chart or map that represents the intersection of the land with the water surface at the elevation of mean lower low water.

mean range of tide (MN) The difference in height between mean high water and mean low water.

mean rise The height of mean high water above the elevation of chart datum.

mean rise interval (MRI) The average interval between the transit of the Moon and the middle of the period of the rise of the tide. It may be computed by adding half the duration of rise to the mean low water interval, subtracting the semi-diurnal tidal period of 12.42 h when greater than this amount. The mean rise interval may be either local or Greenwich according to whether it is referred to the local or Greenwich transit.

mean sea level (MSL) A tidal datum. The arithmetic mean of hourly heights observed over the national tidal datum epoch. Shorter series are specified in the name; for example, monthly mean sea level and yearly mean sea level.

mean sun A fictitious sun that is assumed to move in the celestial equator at a uniform speed corresponding to the average angular speed of the real Sun in the ecliptic, the mean sun being alternately in advance and behind the real Sun. It is used as a reference for reckoning mean time, noon of mean local time corresponding to the time of the transit of the mean sun over the local meridian. See equation of time and mean time.

mean tide level (MTL) A tidal datum. The arithmetic mean of mean high water and mean low water.

mean time Time based upon the hour angle of the mean sun as distinguished from the apparent time which is based upon the position of the real Sun. The difference between apparent and mean time is known as the equation of time.

mean water level (MWL) A datum. The mean surface elevation as determined by averaging the heights of the water at equal intervals of time, usually hourly. Mean water level is used in areas of little or no range in tide.

mean water level line (MWLL) The line on a chart or map that represents the intersection of the land with the water surface at the elevation of mean water level.

meteorological tides Tidal constituents having their origin in the daily or seasonal variations in weather conditions that may occur with some degree of periodicity. The principal meteorological constituents recognized in the tides are S_a, S_{sa}, and S_1. See storm surge.

Metonic cycle	A period of almost 19 years or 235 lunations. Devised by Meton, an Athenian astronomer who lived in the fifth century BC, for the purpose of obtaining a period in which new and full Moon would recur on the same day of the year. Taking the Julian year of 365.25 days and the synodical month as 29.530588 days, we have the 19 year period of 6939.75 days, as compared with the 235 lunations of 6939.69 days, a difference of only 0.06 days.
M_f	Lunar fortnightly constituent. This constituent expresses the effect of departure from a sinusoidal declinational motion:

$$\text{Speed} = 2s = 1.0980331° \text{ per solar hour}$$

mid-extreme tide	An elevation midway between extreme high water and extreme low water, occurring in any locality.
mixed (current)	Type of tidal current characterized by a conspicuous diurnal inequality in the greater and lesser flood strengths and/or greater and lesser ebb strengths. See flood current and ebb current.
mixed (tide)	Type of tide characterized by a conspicuous diurnal inequality in the higher high and lower high waters and/or higher low and lower low waters.
M_m	Lunar monthly constituent. This constituent expresses the effect of irregularities in the Moon's rate of change in distance and speed in orbit:

$$\text{Speed} = s - p = 0.5443747° \text{ per solar hour}$$

modified-range ratio method	A tidal datum computation method. Generally used for the East Coast, Gulf Coast, and Caribbean Island stations. Values needed are mean tide level (MTL), mean diurnal tide level (DTL), mean range of tide (MN), and great diurnal range (GT), as determined by comparison with an appropriate control. From those, the following are computed:

$$\text{MLW} = \text{MTL} - (0.5^*\text{MN})$$

$$\text{MHW} = \text{MLW} + \text{MN}$$

$$\text{MLLW} = \text{DTL} - (0.5^*\text{GT})$$

$$\text{MHHW} = \text{MLLW} + \text{GT}$$

month	The period of the revolution of the Moon around the Earth. The month is designated as sidereal, tropical, anomalistic, nodical, or synodical according to whether the revolution is relative to a fixed star, vernal equinox, perigee, ascending node, or Sun. The calendar month is a rough approximation to the synodical month.
M_{Sf}	Lunisolar synodic fortnightly constituent:

$$\text{Speed} = 2s - 2h = 1.0158958° \text{ per solar hour}$$

mu	Variational constituent. See lambda.

$$\text{Speed} = 2T - 4s + 4h = 27.9682084° \text{ per solar hour}$$

N   Rate of change (as of 1 January 1900) in mean longitude of the
                      Moon's node:

$$N = -0.00220641° \text{ per solar hour}$$

N_2 Larger lunar elliptic semi-diurnal constituent. See L_2:

$$\text{Speed} = 2T - 3s + 2h + p$$

$$= 28.4397295¼ \text{ per solar hour}$$

$2N_2$ Lunar elliptic semi-diurnal second-order constituent:

$$\text{Speed} = 2T - 4s + 2h + 2p$$

$$= 27.8953548¼ \text{ per solar hour}$$

National   A fixed reference adopted as a standard geodetic datum for eleva-
geodetic   tions determined by levelling. The datum was derived for surveys
vertical datum   from a general adjustment of the first-order levelling nets of both
of 1929   the United States and Canada. In the adjustment, mean sea level
(NGVD 1929)   was held fixed as observed at 21 tide stations in the United States
                      and five tide stations in Canada. The year indicates the time of
                      the general adjustment. A synonym for sea level datum of 1929.
                      The geodetic datum is fixed and does not take into account the
                      changing stands of sea level. Because there are many variables
                      affecting sea level, and because the geodetic datum represents a
                      best fit over a broad area, the relationship between the geode-
                      tic datum and local mean sea level is not consistent from one
                      location to another in either time or space. For this reason, the
                      national geodetic vertical datum should not be confused with
                      mean sea level.

National tidal   The specific 19 year period adopted by the National Ocean Ser-
datum epoch   vice as the official time segment over which tide observations are
                      taken and reduced to obtain mean values (e.g. mean lower low
                      water, etc.) for tidal datums. It is necessary for standardization
                      because of periodic and apparent secular trends in sea level. The
                      present national tidal datum epoch is 1960 to 1978. It is reviewed
                      annually for possible revision, and must be actively considered for
                      revision every 25 years.

Neap tides or   Tides of decreased range or tidal currents of decreased speed
tidal currents   occurring semi-monthly as the result of the Moon being in
                      quadrature. The Neap range (Np) of the tide is the average range
                      occurring at the time of Neap tides and is most conveniently com-
                      puted from the harmonic constants. It is smaller than the mean
                      range where the type of tide is either semi-diurnal or mixed, and
                      is of no practical significance where the type of tide is predomi-
                      nantly diurnal. The average height of the high waters of the Neap
                      tide is called Neap high water or high water Neaps (MHWN),
                      and the average height of the corresponding low waters is called
                      Neap low water or low water Neaps (MLWN).

nodal line	A line in an oscillating body of water along which there is a minimum or zero rise and fall of the tide.
nodal point	The zero tide point in an amphidromic region.
node cycle	Period of approximately 18.61 Julian years required for the regression of the Moon's nodes to complete a circuit of 360° of longitude. It is accompanied with a corresponding cycle of changing inclination of the Moon's orbit relative to the plane of the Earth's equator, with resulting inequalities in the rise and fall of the tide and speed of the tidal current.
node factor (f)	A factor depending upon the longitude of the Moon's node that, when applied to the mean coefficient of a tidal constituent, will adapt the same to a particular year for which predictions are to be made.
nodical month	Average period of the revolution of the Moon around the Earth with respect to the Moon's ascending node. It is approximately 27.212220 days in length.
non-harmonic constants	Tidal constants such as lunitidal intervals, ranges, and inequalities that may be derived directly from high and low water observations without regard to the harmonic constituents of the tide. Also applicable to tidal currents.
nu (<2)	Larger lunar evectional constituent:

$$\text{Speed} = 2T - 3s + 4h - p$$

$$= 28.5125831¼ \text{ per solar hour}$$

O_1	Lunar diurnal constituent. See K_1:

$$\text{Speed} = T - 2s + h = 13.9430356° \text{ per solar hour}$$

obliquity factor	A factor in an expression for a constituent tide (or tidal current) involving the angle of inclination of the Moon's orbit to the plane of the Earth's equator.
obliquity of the ecliptic	The angle that the ecliptic makes with the plane of the Earth's equator. Its value is approximately 23.45°.
obliquity of the Moon's orbit	The angle that the Moon's orbit makes with the plane of the Earth's equator. Its value varies from 18.3 to 28.6°, depending upon the longitude of the Moon's ascending node, the smaller value corresponding to a longitude of 180°, and the larger one to a longitude of 0°.
O_{OI}	Lunar diurnal, second-order, constituent:

$$\text{Speed} = T + 2s + h = 16.1391017° \text{ per solar hour}$$

overtide	A harmonic tidal (or tidal-current) constituent with a speed that is an exact multiple of the speed of one of the fundamental constituents derived from the development of the tide-producing force. The presence of overtides is usually attributed to shallow water conditions. The overtides usually considered in tidal

work are the harmonics of the principal lunar and solar semi-diurnal constituents M_2 and S_2, and are designated by the symbols M_4, M_6, M_8, S_4, S_6, etc. The magnitudes of these harmonics relative to those of the fundamental constituents are usually greater in the tidal current than in the tide.

p Rate of change (as of 1 January 1900) in the mean longitude of the lunar perigee:

$$p = 0.00464183° \text{ per solar hour}$$

p_1 Rate of change (as of 1 January 1900) in the mean longitude of the solar perigee:

$$p_1 = 0.00000196° \text{ per solar hour}$$

P_1 Solar diurnal constituent. See K_1:

$$\text{Speed} = T - h = 14.9589314° \text{ per solar hour}$$

parallax   In tidal work, the term refers to the horizontal parallax, which is the angle formed at the centre of a celestial body between a line to the centre of the Earth and a line tangent to the Earth's surface. Since the sine of a small angle is approximately equal to the angle itself in radians, it is usually taken in tidal work simply as the ratio of the mean radius of the Earth to the distance of the tide-producing body. Since the parallax is a function of the distance of a celestial body, the term is applied to tidal inequalities arising from the changing distance of the tide-producing body.

parallax inequality   The variation in the range of tide or in the speed of a tidal current owing to changes in the distance of the Moon from the Earth. The range of tide and speed of the current tend alternately to increase and decrease as the Moon approaches its perigee and apogee respectively, the complete cycle being the anomalistic month. There is a similar but relatively unimportant inequality due to the Sun, the cycle being the anomalistic year. The parallax has little direct effect upon the lunitidal intervals but tends to modify the phase effect. When the Moon is in perigee, the priming and lagging of the tide owing to the phase are diminished, and when in apogee the priming and lagging are increased.

parallax reduction   A processing of observed high and low waters to obtain quantities depending upon changes in the distance of the Moon, such as perigean and apogean ranges.

pelorus   An instrument formerly used on a vessel in connection with a current line and current pole to obtain the set of the current. In its simplest form, it was a disc about 8 inches in diameter and graduated clockwise for every 5 or 10°. It was mounted rigidly on the vessel, usually with the 0° mark forward and the diameter through this mark parallel with the keel. Bearings were then related to the vessel's compass and converted to true.

perigean tides or tidal currents	Tides of increased range or tidal currents of increased speed occurring monthly as a result of the Moon being in perigee. The perigean range (Pn) of tide is the average range occurring at the time of perigean tides and is most conveniently computed from the harmonic constants. It is larger than the mean range where the type of tide is either semi-diurnal or mixed, and is of no practical significance where the type of tide is predominantly diurnal.
perigee	The point in the orbit of the Moon or a man-made satellite nearest to the Earth. The point in the orbit of a satellite nearest to its companion body.
perihelion	The point in the orbit of the Earth (or other planet, etc.) nearest to the Sun.
period	Interval required for the completion of a recurring event, such as the revolution of a celestial body, **or** the time between two consecutive like phases of the tide or tidal current. A period may be expressed in angular measure as 360°. The word is also used to express any specified duration of time.
permanent current	A current that runs continuously and is independent of tides and other temporary causes. Permanent currents include the general surface circulation of the oceans.
phase	Any recurring aspect of a periodic phenomenon, such as new Moon, high water, flood strength, etc. A particular instant of a periodic function expressed in angular measure and reckoned from the time of its maximum value, the entire period of the function being 360°. The maximum and minimum of a harmonic constituent have phase values of 0 and 180° respectively.
phase inequality	Variations in the tides or tidal currents owing to changes in the phase of the Moon. At the times of new and full Moon, the tide-producing forces of the Moon and Sun act in conjunction, causing the range of tide and speed of the tidal current to be greater than the average, the tides at these times being known as Spring tides. At the times of the quadratures of the Moon, these forces are opposed to each other, causing Neap tides with diminished range and current speed.
phase reduction	A processing of observed high and low waters to obtain quantities depending upon the phase of the Moon, such as the Spring and Neap ranges of tide. At a former time this process was known as second reduction. Also applicable to tidal currents.
potential, tide-producing	Tendency for particles on the Earth to change their positions as a result of the gravitational interactions between the Sun, Moon, and Earth. Although gravitational attraction varies inversely as the square of the distance of the tide-producing body, the resulting potential varies inversely as the cube of the distance.
pressure gradient force, horizontal	The horizontal component of the product of the specific volume and the rate of decrease in pressure with distance.

pressure sensor A pressure transducer sensing device for water level measurement. A relative transducer is vented to the atmosphere and pressure readings are made relative to atmospheric pressure. An absolute transducer measures the pressure at its location. The readings are then corrected for barometric pressure taken at the surface.

primary control tide station A tide station at which continuous observations have been made over a minimum of 19 years. Its purpose is to provide data for computing accepted values of the harmonic and non-harmonic constants essential to tide predictions and to the determination of tidal datums for charting and for coastal and marine boundaries. The data series from this station serves as a primary control for the reduction of relatively short series from subordinate tide stations through the method of comparison of simultaneous observations and for monitoring long-period sea level trends and variations. See tide station, secondary control tide station, tertiary tide station, and subordinate tide station.

prime meridian The meridian of longitude which passes through the original site of the Royal Observatory in Greenwich, England, and is used as the origin of longitude. Also known as the Greenwich meridian.

priming of tide The periodic acceleration in the time of occurrence of high and low waters owing to changes in the relative positions of the Sun and Moon.

progressive wave A wave that advances in distance along the sea surface or at some intermediate depth. Although the waveform itself travels significant distances, the water particles that make up the wave merely describe circular (in relatively deep water) or elliptical (in relatively shallow water) orbits. With high, steep, wind waves, a small overlap in the orbital motion becomes significant. This overlapping gives rise to a small net mass transport. See long-shore current and rip current. Progressive waves can be internal, travelling along a sharp density discontinuity, such as the thermocline, or in a layer of gradually changing density (vertically).

pycnocline A layer in which the density increases significantly (relative to the layers above and below) with depth.

Q_1 Larger lunar elliptic diurnal constituent. See M_1:

$$\text{Speed} = T - 3s + h + p$$
$$= 13.3986609\tfrac{1}{4} \text{ per solar hour}$$

$2Q_1$ Lunar elliptic diurnal, second-order, constituent:

$$\text{Speed} = T - 4s + h + 2p$$
$$= 12.8542862\tfrac{1}{4} \text{ per solar hour}$$

quadrature of Moon Position of the Moon when its longitude differs by 90° from the longitude of the Sun. The corresponding phases are known as first quarter and last quarter.

R_2 Smaller solar elliptic constituent. This constituent, with T_2, modulates the amplitude and frequency of S_2 for the effect of variation in the Earth's orbital speed owing to its elliptical orbit:

$$\text{Speed} = 2T + h - p_1 = 30.0410667° \text{ per solar hour}$$

race   A very rapid current through a comparatively narrow channel.

radiational tide   Periodic variations in sea level primarily related to meteorological changes such as the semi-daily (solar) cycle in barometric pressure, daily (solar) land and sea breezes, and seasonal (annual) changes in temperature. Other sea level changes due to meteorological changes that are random in phase are not considered radiational tides.

range of tide   The difference in height between consecutive high and low waters. The mean range is the difference in height between mean high water and mean low water. The great diurnal range or diurnal range is the difference in height between mean higher high water and mean lower low water. For other ranges, see Spring, Neap, perigean, apogean, and tropic tides, and tropic ranges.

real-time   Pertains to a data-collecting system that monitors an ongoing process and disseminates measured values before they are expected to have changed significantly.

reduction factor (F)   Reciprocal of the node factor (f).

reduction of tides or tidal currents   A processing of observed tide or tidal-current data to obtain mean values for tidal or tidal-current constants.

reference station   A tide or current station for which independent daily predictions are given in the tide tables and tidal-current tables, and from which corresponding predictions are obtained for subordinate stations by means of differences and ratios.

relative mean sea level change   A local change in mean sea level relative to a network of benchmarks established in the most stable and permanent material available (bedrock, if possible) on the land adjacent to the tide station location. A change in relative mean sea level may be composed of both an absolute mean sea level change component and a vertical land movement change component.

residual current   The observed current minus the astronomical tidal current.

response analysis   For any linear system, an input function $X_i(t)$ and an output function $X_o(t)$ can be related according to the formula

$$X_o(t) = o_{14} X_i(t - J) \, W(J) \, dJ + \text{noise}(t)$$

where $W(J)$ is the impulse response of the system and its Fourier transform:

$$Z(f) = o_{14} W(J) \, e - 2B_i fJ = R(f) \, e_i N(f)$$

$Z(f)$ is the system's admittance (coherent output/input) at frequency f. In practice, the integrals are replaced by summations; X_i, W, and Z are generally complex. The discrete set of W values are termed response weights, $X_o(t)$ is ordinarily an observed tidal time series, and $X_i(t)$ is the tide potential or the tide at some nearby place. A future prediction can be prepared by applying the weights to an appropriate $X_i(t)$ series. In general:

$$^*Z^* = R(f) \text{ and } \tan(Z) = N(f)$$

measure the relative magnification and phase lead of the station at frequency f.

reversing current   A tidal current that flows alternately in approximately opposite directions with a slack water at each reversal of direction. Currents of this type usually occur in rivers and straits where the direction of flow is more or less restricted to certain channels. When the movement is towards the shore or up a stream, the current is said to be flooding, and when in the opposite direction it is said to be ebbing. The combined flood and ebb movement (including the slack water) covers, on average, 12.42 h for a semidiurnal current. If unaffected by a non-tidal flow, the flood and ebb movements will each last about 6 h, but when combined with such a flow the durations of flood and ebb may be quite different. During the flow in each direction, the speed of the current will vary from zero at the time of slack water to a maximum about midway between the slacks.

rho   Larger lunar evectional diurnal constituent:

$$\text{Speed} = T - 3s + 3h - p$$
$$= 13.4715145\tfrac{1}{4} \text{ per solar hour}$$

rip   Agitation of water caused by the meeting of currents or by a rapid current setting over an irregular bottom. Termed tide rip when a tidal current is involved.

river current   The gravity-induced seaward flow of fresh water originating from the drainage basin of a river. In the freshwater portion of the river below head of tide, the river current is alternately increased and decreased by the effect of the tidal current. After entering a tidal estuary, river current is the depth-averaged mean flow through any cross-section.

rotary current   A tidal current that flows continually, with the direction of flow changing through all points of the compass during the tidal period. Rotary currents are usually found offshore where the direction of flow is not restricted by any barriers. The tendency for the rotation in direction has its origin in the Coriolis force, and, unless modified by local conditions, the change is clockwise in the Northern Hemisphere and counterclockwise in the Southern Hemisphere. The speed of the current usually varies throughout the tidal cycle, passing through the two maxima in approximately opposite directions and through the two minima with the direction of the current at approximately 90° from the directions of the maxima.

s
Rate of change (as of 1 January 1900) in the mean longitude of the Moon:

$$s = 0.54901653° \text{ per solar hour}$$

S_1
Solar diurnal constituent:

$$\text{Speed} = T = 15.0000000° \text{ per solar hour}$$

S_2
Principal solar semi-diurnal constituent. This constituent represents the rotation of the Earth with respect to the Sun:

$$\text{Speed} = 2T = 30.0000000° \text{ per solar hour}$$

S_4, S_6
Shallow water overtides of the principal solar constituent:

$$\text{Speed of } S_4 = 2S_2 = 4T$$
$$= 60.0000000¼ \text{ per solar hour}$$
$$\text{Speed of } S_6 = 3S_2 = 6T$$
$$= 90.0000000¼ \text{ per solar hour}$$

S_a
Solar annual constituent. This constituent, with S_{sa}, accounts for the non-uniform changes in the Sun's declination and distance. In actuality, they mostly reflect yearly meteorological variations influencing sea level:

$$\text{Speed} = h = 0.04106864° \text{ per solar hour}$$

S_{sa}
Solar semi-annual constituent. See S_a:

$$\text{Speed} = 2h = 0.0821373° \text{ per solar hour}$$

salinity (S)
The total amount of solid material, in grams, contained in 1 kg of sea water when all the carbonate has been converted to oxide, the bromine and iodine replaced by chlorine, and all organic matter completely oxidized. The following is approximate:

$$S(‰) = 1.80655 × Cl(‰)$$

where $Cl(‰)$ is chlorinity in parts per thousand.

Saros
A period of 223 synodic months corresponding approximately to 19 eclipse years or 18.03 Julian years. It is a cycle in which solar and lunar eclipses repeat themselves under approximately the same conditions.

secular trend
See apparent secular trend as the preferred term.

seiche   A stationary wave usually caused by strong winds and/or changes in barometric pressure. It is found in lakes, semi-enclosed bodies of water, and in areas of the open ocean. The period of a seiche in an enclosed rectangular body of water is usually represented by the formula

$$\text{Period (T)} = 2L/\%gd$$

in which L is the length, d the average depth of the body of water, and g the acceleration of gravity. See standing wave.

semi-diurnal   Having a period or cycle of approximately one-half of a tidal day. The predominant type of tide throughout the world is semi-diurnal, with two high waters and two low waters each tidal day. The tidal current is said to be semi-diurnal when there are two flood and two ebb periods each day. A semi-diurnal constituent has two maxima and two minima each constituent day, and its symbol is the subscript 2.

sequence of current   The order of occurrence of the four tidal-current strengths of a day, with special reference as to whether the greater flood immediately precedes or follows the greater ebb.

sequence of tide   The order in which the four tides of a day occur, with special reference as to whether the higher high water immediately precedes or follows the lower low water.

set (of current)   The direction towards which the current flows.

shallow water constituent   A short-period harmonic term introduced into the formula of tidal (or tidal-current) constituents to account for the change in the form of a tide wave resulting from shallow water conditions. Shallow water constituents include the overtides and compound tides.

shallow water wave   A wave is classified as a shallow water wave whenever the ratio of the depth (the vertical distance of the still water level from the bottom) to the wave length (the horizontal distance between crests) is less than 0.04. Such waves propagate according to the formula

$$C = \sqrt{gd}$$

where C is the wave speed, g the acceleration of gravity, and d the depth. Tidal waves are shallow water waves.

shear   A quasi-horizontal layer moving at a different velocity relative to the layer directly below and/or above.

sidereal day   The time of the rotation of the Earth with respect to the vernal equinox. It equals approximately 0.99727 of a mean solar day. Because of the precession of the equinoxes, the sidereal day thus defined is slightly less than the period of rotation with respect to the fixed stars, but the difference is less than a hundredth part of a second.

sidereal month   Average period of the revolution of the Moon around the Earth with respect to a fixed star, equal to 27.321661 mean solar days.

sidereal time	This is usually defined by astronomers as the hour angle of the vernal equinox. The sidereal day is the interval between two successive upper transits of the vernal equinox. It is to be noted that, when applied to the month and year, the word sidereal has reference to motion with respect to the fixed stars, while the word tropical is used for motion with respect to the vernal equinox. Because of the precession of the equinox, there is a slight difference.
sidereal year	Average period of the revolution of the Earth around the Sun with respect to a fixed star. Its length is approximately 365.2564 mean solar days.
slack; ebb begins (slack before ebb)	The slack water immediately preceding the ebb current.
slack; flood begins (slack before flood)	The slack water immediately preceding the flood current.
slack water (slack)	The state of a tidal current when its speed is near zero, especially the moment when a reversing current changes direction and its speed is zero. The term is also applied to the entire period of low speed near the time of turning of the current when it is too weak to be of any practical importance in navigation. The relation of the time of slack water to the tidal phases varies in different localities. For a perfect standing tidal wave, slack water occurs at the time of high and of low water, while, for a perfect progressive tidal wave, slack water occurs midway between high and low water.
small diurnal range (Sl)	Difference in height between mean lower high water and mean higher low water.
small tropic range (Sc)	Difference in height between tropic lower high water and tropic higher low water.
solar day	The period of the rotation of the Earth with respect to the Sun. The mean solar day is the time of the rotation with respect to the mean sun. The solar day commencing at midnight is called a civil or calendar day, but, if the day is reckoned from noon, it is known as an astronomical day because of its former use in astronomical calculation.
solar tide	The part of the tide that is due to the tide-producing force of the Sun. The observed tide in areas where the solar tide is dominant. This condition provides for phase repetition at about the same time each solar day.
solar time	Time measured by the hour angle of the Sun. It is called the apparent time when referred to the actual Sun and mean time when referred to the mean sun. It is also classified as local, standard, or Greenwich according to whether it is reckoned from the local, standard, or Greenwich meridian.
solitary wave	A wave of translation consisting of a single crest rising above the undisturbed water level without any accompanying trough. The rate of advance of a solitary wave

depends upon the depth of the water and is usually expressed by the formula

$$C = \sqrt{g(yd + h)}$$

in which C = rate of advance, g = acceleration of gravity, d = depth of water, and h = height of wave, the depth and height being measured from the undisturbed water level.

solstices
: The two points in the ecliptic where the Sun reaches its maximum and minimum declinations; also the times when the Sun reaches these points. The maximum north declination occurs on or near 21 June, marking the beginning of summer in the Northern Hemisphere and the beginning of winter in the Southern Hemisphere. The maximum south declination occurs on or near 22 December, marking the beginning of winter in the Northern Hemisphere and the beginning of summer in the Southern Hemisphere.

Solstitial tides
: Tides occurring near the times of the solstices. The tropic range may be expected to be especially large at these times.

species of constituent
: A classification depending upon the period of a constituent. The principal species are semi-diurnal, diurnal, and long period.

specific volume anomaly, or steric anomaly
: The excess in specific volume over the standard specific volume at 35‰, 0 °C, and the given pressure.

speed (of constituent)
: The rate of change in the phase of a constituent, usually expressed in deg h^{-1}. The speed is equal to 360° divided by the constituent period expressed in hours.

speed (of current)
: The magnitude of velocity. Rate at which the current flows. Usually expressed in knots or cm s^{-1}.

Spring tides or tidal currents
: Tides of increased range or tidal currents of increased speed occurring semi-monthly as a result of the Moon being new or full. The Spring range (Sg) of tide is the average range occurring at the time of Spring tides and is most conveniently computed from the harmonic constants. It is larger than the mean range where the type of tide is either semi-diurnal or mixed, and is of no practical significance where the type of tide is predominantly diurnal. The average height of the high waters of the Spring tides is called Spring high water or mean high water Springs (MHWS), and the average height of the corresponding low waters is called Spring low water or mean low water Springs (MLWS).

stand of tide
: Sometimes called a platform tide. An interval at high or low water when there is no sensible change in the height of the tide. The water level is stationary at high and low water for only an instant, but the change in level near these times is so slow that it is not usually perceptible. In general, the duration of the apparent stand will depend upon the range of tide, being longer for a small range than for a large range, but, where there is a tendency for a double tide, the stand may last for several hours even with a large range of tide.

standard method	A tidal datum computation method. Generally used for the West Coast and Pacific Island stations. Values needed are mean tide level (MTL), mean range of tide (MN), great diurnal range (GT), and mean diurnal high and low water inequalities (DHQ and DLQ), as determined by comparison with an appropriate control. From those, the following are computed:

$$MLW = MTL - (0.5^*MN)$$

$$MHW = MLW + MN$$

$$MLLW = MLW - DLQ$$

$$MHHW = MHW + DHQ$$

standard time	A kind of time based upon the transit of the Sun over a certain specified meridian, called the time meridian, and adopted for use over a considerable area. With a few exceptions, standard time is based upon some meridian that differs by a multiple of $15°$ from the meridian of Greenwich. The United States first adopted standard time in 1883 on the initiative of the American Railway Association, and at noon on 18 November of that year the telegraphic time signals from the Naval Observatory at Washington were changed to this system.
standing (stationary) wave	A wave that oscillates without progressing. One-half of such a wave may be illustrated by the oscillation of the water in a pan that has been tilted. Near the axis, which is called the node or nodal line, there is no vertical rise and fall of the water. The ends of the wave are called loops, and at these places the vertical rise and fall is at a maximum. The current is a maximum near the node and minimum at the loops. A stationary wave may be resolved into two progressive waves of equal amplitude and equal speeds moving in opposite directions.
stationary wave theory	An assumption that the basic tidal movement in the open ocean consists of a system of stationary wave oscillations, any progressive wave movement being of secondary importance except as the tide advances into tributary waters. The continental masses divide the sea into irregular basins, which, although not completely enclosed, are capable of sustaining oscillations that are more or less independent. The tide-producing force consists principally of two parts, a semi-diurnal force with a period of approximately half a day and a diurnal force with a period of approximately a whole day. Insofar as the free period of oscillation of any part of the ocean, as determined by its dimensions and depth, is in accord with the semi-diurnal or diurnal tide-producing forces, there will be built up corresponding oscillations of considerable amplitude that will be manifested in the rise and fall of the tide. The diurnal oscillations, superimposed upon the semi-diurnal oscillations, cause the inequalities in the heights of the two high and the two low waters of each day. Although the tidal movement as a whole

is somewhat complicated by the overlapping of oscillating areas, the theory is consistent with observational data.

storm surge The local change in the elevation of the ocean along a shore due to a storm. The storm surge is measured by subtracting the astronomic tidal elevation from the total elevation. It typically has a duration of a few hours. Since wind-generated waves ride on top of the storm surge (and are not included in the definition), the total instantaneous elevation may greatly exceed the predicted storm surge plus astronomic tide. It is potentially catastrophic, especially on low-lying coasts with gently sloping offshore topography.

storm tide As used by the National Weather Service (NOAA), the sum of the storm surge and astronomic tide. See storm surge.

subordinate current station (1) A current station from which a relatively short series of observations is reduced by comparison with simultaneous observations from a control current station. See current station, control current station, and reference station. (2) A station listed in the tidal-current tables for which predictions are to be obtained by means of differences and ratios applied to the full predictions at a reference station. See reference station.

subordinate tide station (1) A tide station from which a relatively short series of observations is reduced by comparison with simultaneous observations from a tide station with a relatively long series of observations. See tide station, primary control tide station, secondary control tide station, and tertiary tide station. (2) A station listed in the Tide Tables from which predictions are to be obtained by means of differences and ratios applied to the full predictions at a reference station. See reference station.

summer time British name for daylight saving time.

synodical month The average period of the revolution of the Moon around the Earth with respect to the Sun, **or** the average interval between corresponding phases of the Moon. The synodical month is approximately 29.530588 days in length.

syzygy With respect to tides, whenever the Moon is lined up with the Earth and Sun in a straight Sun–Moon–Earth or Sun–Earth–Moon configuration. At these times, the range of tide is greater than average.

T Rate of change in hour angle of mean sun at place of observation:

$$T = 15° \text{ per mean solar hour}$$

T_2 Larger solar elliptic constituent. See R_2:

$$\text{Speed} = 2T - h + p_1 = 29.9589333° \text{ per solar hour}$$

terdiurnal Having three periods in a constituent day. The symbol of a terdiurnal constituent is the subscript.

tidal bore A tidal wave that propagates up a relatively shallow and sloping estuary or river with a steep wave front. The leading edge presents an abrupt rise in level, frequently with continuous breaking and often immediately followed by several large undulations. An uncommon phenomenon, the tidal bore is usually associated with very large ranges in tide as well as wedge-shaped and rapidly shoaling entrances. Also called eagre, eager (for Tsientan, China bore), mascaret (French), pororoca (Brazilian), and bore.

tidal characteristics Principally, those features relating to the time, range, and type of tide.

tidal constants Tidal relations that remain practically constant for any particular locality. Tidal constants are classified as harmonic and non-harmonic. The harmonic constants consist of the amplitudes and epochs of the harmonic constituents, and the non-harmonic constants include the ranges and intervals derived directly from the high and low water observations.

tidal current A horizontal movement of the water caused by gravitational interactions between the Sun, Moon, and Earth. The horizontal component of the particulate motion of a tidal wave. Part of the same general movement of the sea that is manifested in the vertical rise and fall called tide.

Tidal-current chart diagrams A series of monthly diagrams to be used with the tidal current charts. Each diagram contains lines that indicate the specific tidal-current chart to use for a given date and time, and the speed factor to apply to that chart.

Tidal-current charts Charts on which tidal-current data are depicted. Tidal-current charts for a number of important waterways are published by the National Ocean Service in the United States and by the British Admiralty. Each consists of a set of charts giving the speed and direction of the current for each hour or equal interval of the tidal cycle, thus presenting a comprehensive view of the tidal-current movement.

Tidal-current tables Tables that give daily predictions of the times and velocities of the tidal currents. These predictions are usually supplemented by current differences and constants through which predictions can be obtained for numerous other locations.

tidal difference Difference in time or height between a high or low water at a subordinate station and a reference station for which predictions are given in the tide tables. The difference, when applied according to sign to the prediction at the reference station, gives the corresponding time or height for the subordinate station.

tidal stream British equivalent of United States tidal current.

tidal wave A shallow water wave caused by the gravitational interactions between the Sun, Moon, and Earth. Essentially, high water is the crest of a tidal wave, and low water is the trough. Tidal current is the horizontal component of the particulate motion, while tide is manifested by the vertical component. The observed tide and tidal current can be considered the result of the combination of several tidal waves, each of which may vary from nearly pure progressive to nearly pure standing, and with differing periods, heights, phase relationships, and direction.

tide

The periodic rise and fall of a body of water resulting from gravitational interactions between the Sun, Moon, and Earth. The vertical component of the particulate motion of a tidal wave. Although the accompanying horizontal movement of the water is part of the same phenomenon, it is preferable to designate this motion as tidal current.

tide curve

A graphic representation of the rise and fall of the tide in which time is usually represented by the abscissa and height by the ordinate. For a semi-diurnal tide with little diurnal inequality, the graphic representation approximates a cosine curve. See marigram.

tide-predicting machine

A mechanical analogue machine especially designed to handle the great quantity of constituent summations required in the harmonic method.

tide-producing force

That part of the gravitational attraction of the Moon and Sun that is effective in producing the tides on the Earth. The force varies approximately as the mass of the attracting body and inversely as the cube of its distance. The tide-producing force exerted by the Sun is a little less than one-half as great as that of the Moon.

tide (water level) station

The geographic location at which tidal observations are conducted. Also, the facilities used to make tidal observations. These may include a tide house, tide (water level) gauge, tide staff, and tidal benchmarks. See primary control tide station, secondary control tide station, tertiary tide station, and subordinate tide station.

Tide tables

Tables that give daily predictions of the times and heights of high and low waters. These predictions are usually supplemented by tidal differences and constants through which predictions can be obtained for numerous other locations.

tideway

A channel through which a tidal current flows.

time

Time is measured by the rotation of the Earth with respect to some point in the celestial sphere and may be designated as sidereal, solar, or lunar, according to whether the measurement is taken in reference to the vernal equinox, the Sun, or the Moon. Solar time may be apparent or mean, according to whether the reference is to the actual Sun or the mean sun. Mean solar time may be local or standard, according to whether it is based upon the transit of the Sun over the local meridian or a selected meridian adopted as a standard over a considerable area. Greenwich time is standard time based upon the meridian of Greenwich. In civil time, the day commences at midnight, while in astronomical time, as used prior to 1925, the beginning of the day was reckoned from noon of the civil day of the same date. The name universal time is now applied to Greenwich mean civil time.

time meridian

A meridian used as a reference for time.

total current The combination of the tidal and non-tidal current. The United States equivalent of the British flow.

tractive force The horizontal component of a tide-producing force vector (directed parallel with level surfaces at that geographic location).

transit The passage of a celestial body over a specified meridian. The passage is designated as upper transit or lower transit according to whether it is over that part of the meridian lying above or below the polar axis.

tropic currents Tidal currents occurring semi-monthly when the effect of the Moon's maximum declination is greatest. At these times the tendency of the Moon to produce a diurnal inequality in the current is at a maximum.

tropic inequalities Tropic high water inequality (HWQ) is the average difference between the two high waters of the day at the times of tropic tides. Tropic low water inequality (LWQ) is the average difference between the two low waters of the day at the times of tropic tides. These terms are applicable only when the type of tide is semi-diurnal or mixed.

tropic intervals Tropic higher high water interval (TcHHWI) is the lunitidal interval pertaining to the higher high waters at the time of the tropic tides. Tropic lower low water interval (TcLLWI) is the lunitidal interval pertaining to the lower low waters at the time of the tropic tides. Tropic intervals are marked *a* when reference is made to the upper transit of the Moon at its north declination or to the lower transit at the time of south declination, and are marked *b* when the reference is to the lower transit at the north declination or to the upper transit at the south declination.

tropic ranges The great tropic range (G_c), or tropic range, is the difference in height between tropic higher high water and tropic lower low water. The small tropic range (S_c) is the difference in height between tropic lower high water and tropic higher low water. The mean tropic range (M_c) is the mean between the great tropic and the small tropic range. Tropic ranges are most conveniently computed from the harmonic constants.

tropic speed The greater flood or greater ebb speed at the time of tropic currents.

tropic tides Tides occurring semi-monthly when the effect of the Moon's maximum declination is greatest. At these times there is a tendency for an increase in the diurnal range. The tidal datums pertaining to the tropic tides are designated as tropic higher high water (TcHHW), tropic lower high water (TcLHW), tropic higher low water (TcHLW), and tropic lower low water (TcLLW).

tropical month The average period of the revolution of the Moon around the Earth with respect to the vernal equinox. Its length is approximately 27.321582 days.

tropical year The average period of the revolution of the Earth around the Sun with respect to the vernal equinox. Its length is approximately

365.2422 days. The tropical year determines the cycle of changes in the seasons, and is the unit to which the calendar year is adjusted through the occasional introduction of the extra day on leap years.

trough The lowest point in a propagating wave.

true direction Direction relative to true north (0°) which is the direction of the north geographic pole. See compass direction and magnetic direction.

tsunami A shallow water progressive wave, potentially catastrophic, caused by an underwater earthquake or volcano.

type of tide A classification based on characteristic forms of a tide curve. Qualitatively, when the two high waters and two low waters of each tidal day are approximately equal in height, the tide is said to be semi-diurnal; when there is a relatively large diurnal inequality in the high or low waters or both, it is said to be mixed; and when there is only one high water and one low water in each tidal day, it is said to be diurnal. Quantitatively (after Dietrich), where the amplitude ratio of $K_1 + O_1$ to $M_2 + S_2$ is <0.25, the tide is classified as semi-diurnal; where it is 0.25 to 1.5, the tide is mixed, mainly semi-diurnal; where it is 1.5 to 3.0, the tide is mixed, mainly diurnal; where it is >3.0, the tide is diurnal.

universal time (UT) Same as Greenwich mean time (GMT). See time, kinds.

upwelling An upward flow of subsurface water owing to such causes as surface divergence, offshore wind, and wind drift transport away from shore.

vanishing tide In a predominantly mixed tide with a very large diurnal inequality, the lower high water and higher low water become indistinct (or vanish) at times of extreme declinations.

variation (of compass) Difference between true north as determined by the Earth's axis of rotation and magnetic north as determined by the Earth's magnetism. Variation is designated as east or positive when the magnetic needle is deflected to the east of true north, and as west or negative when the deflection is to the west of true north. The variation changes with time.

variational inequality An inequality in the Moon's motion owing mainly to the tangential component of the Sun's attraction.

velocity (of current) Speed and set of the current.

wave height The vertical distance between crest and trough.

Z_0 Symbol recommended by the International Hydrographic Organization to represent the elevation of mean sea level above chart datum.

Index